FUEL POVERTY

Fuel poverty
from cold homes to affordable warmth

Brenda Boardman

Belhaven Press
(a division of Pinter Publishers)
London and New York

First published in Great Britain in 1991 by Belhaven Press (a division of Pinter Publishers), 25 Floral Street, London WC2E 9DS

British Library Cataloguing in Publication Data
A CIP catalogue record for this book is available from the British Library

ISBN 1 85293 139 6

For enquiries in North America please contact PO Box 197, Irvington, NY 10533

Library of Congress Cataloging in Publication Data

Boardman, Brenda, 1943-
  Fuel poverty: from cold homes to affordable warmth/Brenda Boardman.
    p. cm.
  Includes bibliographical references and index.
  ISBN 1-85293-139-6
  1. Poor-Great Britain-Energy assistance.   I. Title.
HC260.P63B62 1991   90-23113
363.5'83-dc20   CIP

Typeset by Communitype Communications Limited
Printed and bound in Great Britain by Biddles Limited

Contents

List of Figures

List of Tables

Acknowledgements

This book has grown out of a doctoral thesis undertaken at the Science Policy Research Unit, University of Sussex. My main debt is, therefore, to my two supervisors, John Chesshire and Gordon MacKerron, for teaching me the crafts of research and writing. Their immense patience and determined guidance enabled me to complete the thesis and achieve a standard that I hope does them justice. The encouraging criticisms of my two examiners, Professor John Surrey and Professor David Donnison, have helped with subsequent improvements.

Financial assistance has been received from the Joint Committee of the Science and Engineering Research Council and the Economic and Social Research Council and the Royal Institute of British Architects, supplemented by a substantial subsidy from my husband. I shall always be grateful to them for enabling me to undertake this study.

Many specialists from government departments, research and academic institutions, particularly the Science Policy Research Unit, have assisted in my enquiries and given generously of their time and knowledge. I am not naming them individually — some may wish to remain anonymous — but I could not have tackled such a multi-disciplinary approach without some gentle coaching by people already expert in their particular field.

Figures 6.3, 7.1, 7.2, 7.4 and 9.1 are reproduced by courtesy of the Building Research Establishment and by permission of the Controller of HMSO. I am grateful to the Select Committee on Energy for permission to use Figure 7.6 and to the Editor of Energy Policy and Richard R. Barnett (the author) for the use of Figure 9.2.

Through my colleagues from the National Right to Fuel Campaign, the Energy Conservation and Solar Centre and the Public Utilities Access Forum I have learnt, vicariously, of the problems experienced by those who suffer from fuel poverty. This knowledge helps to ensure that the study is based in reality.

The support and encouragement given to me in my work by John, Emily and Giles has meant that research and family life have continued to coexist in harmony, just. I hope that this work and the support of so many people results in a clearer perception of the problems of fuel poverty, better policies and warmer people.

Brenda Boardman
9 Grange Road
Lewes
East Sussex BN7 1TR

Abbreviations

ach	air changes per hour
AMA	Association of Metropolitan Authorities
BRE	Building Research Establishment
BRECSU	Building Research Energy Conservation Support Unit
BREDEM	Building Research Establishment Domestic Energy Model*
CH	Central heating*
CO_2	Carbon dioxide
COWI	Cost of Warmth Index*
CPAG	Child Poverty Action Group
DD	Degree days*
DEN	Domestic Energy Notes
DEn	Department of Energy
DHSS	Department of Health and Social Security (now DSS)
DOE	Department of the Environment
DSS	Department of Social Security (previously DHSS)
DUKES	Digest of United Kingdom Energy Statistics
EEO	Energy Efficiency Office, Department of Energy
ERHA	Estate rate heating addition*
ESW	Exceptionally severe weather*
EWD	Excess winter deaths*
FES	Family Expenditure Survey
FFL	Fossil Fuel Levy
FPI	Fuel Price Index*
GHS	General Household Survey
HA	Heating addition*
HB	Housing Benefit*
HC	House of Commons
HEES	Home Energy Efficiency Scheme
IRH	Individual room heaters
IS	Income support (replaced SB in April 1988)
MIT	Mean internal temperature
MKECI	Milton Keynes Energy Cost Index*
MOHLG	Ministry of Housing and Local Government
MSC	Manpower Services Commission
NACAB	National Association of Citizens Advice Bureaux
NCC	National Consumer Council

NCH	No central heating
NFE	Notional fuel element*
NFHA	National Federation of Housing Associations
NRFC	National Right to Fuel Campaign
Offer	Office of Electricity Regulation
Ofgas	Office of Gas Supply (the gas regulator)
PES	Public electricity supplier
PSBR	Public sector borrowing requirement
RPI	Retail price index*
RRR	Required rate of return
SB	Supplementary benefit*
SSS	Social Security Statistics
WA	Written Answer (in Hansard)

* defined in Glossary

1 Framework

Fuel poverty is most simply defined as the inability to afford adequate warmth in the home (Lewis 1982, p.1) and was first identified as a distinct issue of public concern following the oil crisis in 1973-4. In essence this was because of the effect of very large fuel price rises on low-income households. Since then, when the weather is particularly severe in winter there is media discussion and questions are asked in Parliament about the plight of cold people and the adequacy of government responses — the subject of fuel poverty is in the news again. A short-term palliative may be announced to cope with the immediate crisis, such as the amended exceptionally severe weather allowance of £5 per week introduced in January 1987. At other times of the year, rapid fuel price rises and the number of households disconnected from their gas or electricity supply may receive similar coverage. Thus, the public debate on fuel poverty is essentially intermittent and seasonal, but recurring.

As the name 'fuel poverty' implies, much of the debate has been on the interaction of real fuel price increases and low incomes. Originally, the solution was seen to lie with fuel-pricing policy and then the shift was towards improving income levels, particularly through heating additions (HA). The condition of the housing stock has been gently mentioned at intervals. Therefore, the three government departments of Energy (DEn), Environment (DOE) and Social Security (DSS — previously Health and Social Security) take turns in the spotlight and try to move out as rapidly as possible. Thus, any explanations for fuel poverty that have emanated from government have to be viewed with a certain scepticism: fuel poverty is a large social problem with considerable spending implications for the department unfortunate enough to be left holding the issue. This is reflected in the recent emphasis on denying the existence of fuel poverty, except as an aspect of general poverty. A view stated by Peter Walker when he was Secretary of State for Energy:

> I am afraid that I must take issue with the term 'fuel poverty'. People do not talk of 'clothes poverty' or 'food poverty', and I do not think that it is useful to talk of 'fuel poverty' either. Of course there is far too much poverty around, and the Government is spending huge sums to alleviate it through the social security system. The regeneration of the economy, for which we are all working, will also help. (Energy Action Bulletin, October 1985)

An almost identical statement was made by John Major, when

Parliamentary Under-Secretary of State for Health and Social Security
(Hansard 16 Dec. 1985, col 135). Thus, two ministers closely involved
with the problem chose to stress the indivisibility of poverty and fuel
poverty. If a problem does not exist, there is no need to spend money on
it, so this logic was extended with the abolition of HA in the social security
changes, April 1988, resulting in a government saving of at least £200m
pa. A major objective of this book is to examine whether and, if so, in
what ways fuel poverty is different from poverty and, having made this
assessment, to identify the implications.

This discussion is predicated upon a concern for the hardship and
suffering caused by fuel poverty, a belief in the role of the welfare state
and in helping the most needy. Another underlying premiss is that
people should have freedom of choice, including the freedom to choose to
be warm. This may seem to be a novel interpretation, but it is important to
stress the concept because warmth is a necessity of life. If the only option
is to be cold, because there is insufficient income to purchase adequate
warmth, then the individual is denied freedom of choice.

Energy efficiency

Energy efficiency in the home affects energy consumption, energy costs
and greenhouse gas emissions. Because the concept is central to this
analysis and to avoid confusion with alternative meanings, energy
efficiency is discussed and differentiated from energy conservation. This
is not a full examination of the history of the two terms, merely an
introduction to some of the common meanings, to clarify what is and is
not meant here by the phrase 'energy efficiency'.

The original discussion was about energy conservation, 'loosely, the
act of saving fuel' (Cooper 1982, p.2). The implicit assumption with
energy conservation is that excess energy-related services are being
obtained and fuel economies are possible, for instance your house is
warm and you would be just as happy in a cooler house with extra
clothing on. Therefore, through behavioural changes and an acceptance
of a lower standard of energy service, less energy will be used; no changes
to capital stocks are necessary. Early definitions of energy conservation
were based on technical measures and lacked price or cost effectiveness
dimensions. For these reasons, the initial approach was seen to be too
narrow:

> 'energy conservation should be seen as the adoption of any measure
> that cost-effectively increases benefits relative to the amount or cost of
> energy consumed — in other words 'the more cost efficient use of
> energy'. (ACEC 1983, p.3)

However, as this last quotation demonstrates, there was still confusion as
to whether the efficient use of energy is judged solely in technical terms or
includes an economic dimension as well. It is the latter, the economic
approach, that now prevails:

The Government views its energy pricing policy as central to its campaign for energy efficiency'. (HC 87 1985, p.vii)

This was confirmed by the Secretary of State for Energy when setting up the Energy Efficiency Office (EEO):

the Government is committed to a policy that will ... encourage increased efficiency in energy use, giving consumers as a whole better value for money'. (DEn 1983a, p.1).

The EEO gave this definition the popular name of 'monergy'. The same approach is used here: energy efficiency is measured in terms of the output of energy service per unit of expenditure on fuel, which in the case of heating means the amount of warmth obtained per £ of expenditure. Increased energy efficiency may sometimes result from the substitution of a cheaper fuel for a more expensive one and this possibility is included within the definition. Invariably, to obtain energy efficiency improvements there has to be investment in capital stocks — it is virtually impossible to obtain improved energy efficiency through behavioural changes alone, as there would have to be unused capital equipment already available. Two inexpensive examples are putting lids on saucepans, where not previously used, or heating hot water on the gas cooker rather than with an electric kettle. Throughout, therefore, the discussion is about energy efficiency, the broader term, and not specifically about energy conservation. Importantly, energy efficiency includes two economic dimensions. First, the cost of the fuel used is incorporated into any measure of energy efficiency and secondly, improvements in energy efficiency have to be cost effective, taking the cost of the fuel and the capital expenditure into account. These dimensions are crucial for low-income households.

Energy efficiency of the extraction, production or conversion process upstream of the house is not included in this study, as it is assumed to be reflected in the cost to the consumer, or in future price variations. These are aspects of the more efficient supply of energy, rather than energy efficiency in the demand sector. To distinguish more clearly, the phrases 'more efficient use of energy' or 'end-use efficiency' are used sometimes in the text. The sole concern here is with the efficient use of energy, even if this is shortened to the phrase 'energy efficiency'.

The focus on technical changes in the capital stocks involved in household energy use reflects a major emphasis in energy studies. For instance, the Energy Group at the Science Policy Research Unit of Sussex University has recognized 'the pivotal importance of the capital equipment used in the energy sector' in its research (SPRU 1988, p.9). Because improved energy efficiency can only be achieved through capital investment, another major objective is to examine the impact of capital stocks and the rate of technical change on the energy efficiency of low-income homes and the extent to which these contribute to fuel poverty.

4   FUEL POVERTY

Energy vs. warmth

The analysis of fuel poverty requires a clear differentiation between the
purchase of fuel and the purchase of an energy service, such as warmth.
The purchase of fuel is a relatively straightforward transaction mainly
involving the interplay of prices and incomes. Warmth is a very different
commodity, as its price depends upon the technical characteristics of the
heating system and the building fabric, as well as the fuel used.

Delivered energy is purchased from the supplier and converted into
useful energy by the appliances in the home. The cost of useful energy
depends upon the purchase price of the fuel and the technical efficiency
with which it is converted into a useful form, such as heat or light. When
the product required is warmth, there is a further component to energy
efficiency: the rate at which heat is lost from the building. In order for the
temperature within the home to rise, the heating system has to be capable
of producing warmth at a faster rate than it is being lost from the building,
and the greater the difference between the two rates of flow, the faster the
building will warm up, and the more quickly the system can be turned off
or used intermittently. Thus, the level of thermal insulation provided by
the building fabric is of vital importance in determining the benefit that is
obtained from the warmth produced by the heating system. This point
was recognized by the Select Committee on Energy in their Report on
Energy Conservation in Buildings when they looked at the
transformation of delivered energy into useful energy and: 'with the more
efficient *use* of useful energy by, for instance, reducing its loss in
buildings' (HC 401-1 1982, p.x — emphasis added). The cost of a unit of
warmth in a home is therefore partly determined by two sets of capital
stocks (the home and the heating system).

The three commodities — delivered energy, useful energy and warmth
— are listed in Table 1.1, together with the way different types of energy
efficiency improvements impact on them. All of these interactions are
discussed further in the text.

Table 1.1 Effect of energy efficiency improvements on the cost of delivered
energy, useful energy and warmth

	Cost of delivered energy	Cost of useful energy	Cost of warmth
Price of fuel falls*	lower	lower	lower
Substitute a cheaper fuel	lower	lower	lower
New boiler — same fuel*	no change	lower	lower
Insulation added	no change	no change	lower

Note:
* Same fuel as used to calculate the cost of delivered and useful energy

The recognition that households wish to purchase warmth, rather than fuel, necessitates a considerable extension of the boundaries to the problem of fuel poverty. These additional areas of study occur in several different disciplines, including the physiological need for warmth and criteria for comfort; the technology involved in heating systems and houses; people's understanding of energy use in the home; and an economic model of expenditure on warmth. Thus a multi-disciplinary approach is required to examine the characteristics of warmth as a commodity and to help disentangle the differences between poverty and fuel poverty.

The concern here begins with delivered energy purchased by individual domestic consumers and with the way this is converted into useful energy within the home (or on the estate in the case of district heating). The practice in this text is to assume that fuel poverty applies to all domestic uses of fuel, even though the major manifestation is inadequate heating. References to cold homes, therefore, do not imply that this is the only form of deprivation resulting from fuel poverty, although it may be the most identifiable (it is easier to confirm that a room is cold than it is to define when a household has inadequate hot water). A full examination of the non-heating uses of energy is not undertaken here, though they are another component of fuel poverty, as the National Right to Fuel Campaign (NRFC) imply in their slogan: 'A warm, well-lit home for everyone'.

Cost of warmth and two examples

The cost of a unit of warmth combines with the demand for warmth to give the total household cost. These interactions are linked in the Cost of Warmth Index (COWI — see Glossary). This is a simplified equation developed by the author (Boardman 1984) to show how the seven main factors influence the cost of keeping warm in a house. The benefit of using the COWI is that it provides a useful tool with which to conceptualize the relationship between the different factors. As evidence of the way in which the COWI works and of the way in which the cost of warmth can vary between advantaged and disadvantaged households, the following two examples are given. These are purely illustrative and no other detailed calculations are undertaken using the COWI formula — it is only used to provide a framework.

HOUSEHOLD A — lives in a detached house with a floor area of 96m^2 and floor-ceiling height of 2.5m. The house is well insulated with 100mm loft insulation, cavity-fill and double glazing with heavyweight curtains. Ventilation losses are minimal. The gas condensing boiler is 85 per cent efficient and costs 1.46p per unit of heat (kWh) at the meter — equivalent to 42.9p/therm. The family live in the south-west of England where the average temperature in the winter months is 8.8°C. They spend five hours a day sitting still and a further three hours a day moderately active in the house, requiring an average temperature of 21°C. When the heating is on it is heating the whole house: no heating is on at night.

HOUSEHOLD B — is an elderly person, over 75, living in a three-roomed flat on the first floor of a Victorian terraced house, with another flat above. Only the front room (22m²) is heated: this has a moderate ceiling height (2.8m) and the windows form 40 per cent of the exterior wall. There has been no insulation added, the windows are worn, the fireplace is poorly blocked, so that ventilation losses are three times greater than in Household A. Whilst there are no overall losses through the party wall, floor and ceiling, there is heat loss through the internal partitions to the rest of the unheated flat. The 3kW electric fire is 100 per cent efficient and uses electricity at 6.62p per delivered kWh. The occupier is out of bed, but in the house for thirteen hours a day: eight hours are spent sitting still and five hours moderately active needing an average of 21°C. No heating is on at night. The house is in the south-east where the average temperature in the winter months is 7°C.

	Household A	Household B
Average daily cost of keeping warm throughout the winter, excluding standing charges and fixed costs	£0.35	£1.68

Thus, Household B pays over four times as much per day as Household A, and only manages to heat about a quarter of the space. Whilst these examples may seem extreme, they indicate the wide variation that can occur in the cost of achieving a given level of required warmth. Conventional wisdom would expect the large detached house of a family to be considerably more expensive to keep warm than a small amount of space in a single person's flat on the middle floor of a terraced house. These examples demonstrate the cumulative impact and cost penalties of living in an energy inefficient home.

Fuel prices

Considerable attention has focused on the impact that fuel prices have in creating or exacerbating fuel poverty. An examination of some of the aspects of fuel pricing helps establish what the contribution of fuel-pricing policies is to fuel poverty. The impact of a fuel price increase will depend upon the budget share of that fuel, which will be determined by what other fuels are used and how much of total expenditure goes on fuel. Thus, an all-electric household spending 15 per cent of the weekly budget on fuel is going to be affected by an increase in the price of electricity much more than a family using gas for cooking, space and water heating and whose electricity bill represents 1 per cent of expenditure. The annual change in the real price of a fuel already being used by a consumer, therefore, affects household budgeting practices and real increases cause particular hardship, if there is no, or a delayed, compensatory increase in income. Sudden and large price rises after the first oil crisis were thought to be a main cause of fuel poverty. This is examined further in Chapter 2.

A fuel price increase affects the cost of useful energy and of warmth in all using households at a single point in time (Table 1.1). The importance of these impacts can only be judged in relation to the amount the poor are disadvantaged in the areas covered by other COWI factors. Thus, a household's reaction to a fuel price rise will depend upon both the cost of warmth and the value placed on it (Chapter 9).

As the above examples (Households A and B) illustrate, the relative price of fuels at the same point in time, and thus the variations in the price of warmth, are crucial in creating fuel poverty. The wide range (general tariff electricity is four times the price of gas, at the meter) and the stability of these variations are the main aspects of fuel pricing dealt with in this book.

These two short-term responses to fuel prices and the underlying relative costs are the main focus of the study. The influence of fuel prices on household capital investment decisions, for instance when purchasing a heating system, is beyond the scope of this study, as low-income households rarely have that option. Similarly, little attention is given to the role of fuel over time in the household budget, particularly in comparison with other expenditures. Finally, the reasons for a fuel price change — macro energy policy — are mentioned briefly, but not examined in detail (Chapters 2 and 9). These different elements of fuel prices have been rehearsed to clarify the wide variety of issues encompassed in a single phrase. The past emphasis on indices has tended to obscure the vitally important relative costs.

Global warming

Global warming is inextricably linked to the use of energy, as 80 per cent of all global carbon dioxide (CO_2) emissions come from burning fossil fuel (Darmstadter and Edmonds 1989, p.37) and 27 per cent of UK carbon dioxide comes from domestic energy use. UK and world consumption of energy is increasing and so, therefore, are levels of emissions, with the risk that we are overloading the atmosphere and profoundly altering the climate. The concentration of these gases in the air results in more of the sun's heat being trapped in the lower atmosphere. It is the excess of these 'greenhouse' gases that is linked to global warming. There is considerable uncertainty about the likely effects of global warming and the speed with which they might occur. However, delay now may result in the need for greater and more radical intervention in the near future.

Several gases contribute to the greenhouse effect, but carbon dioxide is the most serious, as it is responsible for half of the atmospheric deterioration. The main domestic fuels vary in the amount of carbon dioxide that they release and in their price to the consumer. The least pollution at the lowest price comes from the use of natural gas, whereas general tariff electricity is by far the most expensive fuel and the most polluting. Policies to reduce the cost of keeping warm in low-income homes will, generally, also have the effect of cutting down on carbon dioxide emissions. The one exception to this concerns off-peak electricity tariffs and is discussed in Chapter 5.

Concerns about global warming and the role of energy efficiency in the reduction of emissions mean that many of the examples and lessons from this study are illustrative of wider issues. The relative values of policies based on revenue issues (taxes, income support) versus those using direct capital investment will be the foundation of many discussions on greenhouse problems. The need is to ensure that low-income households are not further disadvantaged by policies to reduce carbon dioxide emissions.

Structure of the book

A wide range of literature has been used to obtain sufficient information. Despite much detailed analysis this results in what may only be described as a statistical patchwork that varies both in detail and coverage. Large and small-scale surveys, recent and older data have been used, the majority having been collected, in the first instance, for other purposes. Similarly, some important sources are pressure groups and tenants, as well as the more conventional academic and government sources:

Not being able to heat the house; crying with the cold (Family on supplementary benefit, quoted in CPAG 1980, p.4)

The emergence of fuel poverty as a recognized problem, the context in which it occurred and early government responses are discussed in Chapter 2. Some of the background issues, such as the different emphasis placed on space and warmth in housing standards, are sketched in to show the context in which fuel poverty developed. The continuing evidence of fuel poverty confirms that past policy initiatives have not been successful and that both an improved understanding and a clearer policy focus are required.

The similarities between the poor and the fuel poor are considered in Chapter 3. Because of the problems in identifying those that suffer from fuel poverty, different definitions of poverty are reviewed in order to provide a working basis for the analysis contained in the rest of this study. The social characteristics of the poor and of those people suffering from the income-related indicators of fuel poverty, such as debts and disconnections, are compared to assess whether the two groups have common identities. Other social factors that might contribute to fuel poverty, such as education and knowledge, are examined.

The detailed analysis of the factors affecting the cost of a unit of warmth commences in Chapter 4 with an examination of heat retention by the building fabric. The processes involved are heat conduction through the fabric (COWI 1) and ventilation losses (COWI 2). These are examined separately to emphasize their relative importance. The variations that occur between households are discussed and the extent to which government policies are widening the range.

The energy efficiency of the heating system inside the building is the other component of the cost of a unit of warmth (Chapter 5). The cost of useful energy is derived from the technical efficiency of the appliance

(COWI 3) and the cost of the fuel used (COWI 4). The variations between households and how these have altered over time are examined, together with the government's role.

The total cost of keeping warm depends both on the price of each unit of warmth and on the quantity demanded. This quantity is a product of temperature, time and space. Previously, most attention has been given to temperature levels and there has been little discussion on the other two components. The absence of a workable definition of adequate warmth is symptomatic of the low priority which in the past has been given to the provision of warmth in building and environmental health standards. To clarify the issues, a heating standard has been formulated (Chapter 6) upon which can be based a definition of adequate warmth and, thus, fuel poverty. The data used in defining this heating standard include evidence of the temperatures required for comfort and health, when there is no economic constraint (COWI 6). From this is subtracted the external temperature, as modified by non-heating incidental gains (COWI 5) to give the temperature gain to be achieved through heating. The hours of heating and the proportion of the house to be heated (COWI 7) complete the elements in the heating standard. None of these last three COWI factors can be directly influenced by government policy.

The relationship between the target heating standard and actual measured conditions is discussed in Chapter 7 to investigate the extent of cold homes and to identify influential factors. Because of the paucity of measured data, Britain is compared with other countries to explore the relationship between temperatures inside and outside the house, and deaths in winter.

It is not possible to draw a neat line around energy consumption for heating and isolate this from expenditure on other energy services. This means that although the primary focus is on the heating aspects of fuel poverty, other energy uses have to be considered when energy expenditure is examined (Chapter 8). However, all uses of energy in the home result in some incidental heat output, for instance from cooking, lighting, refrigeration and hot water, and thus reduce the need for space heating, so that total energy consumption is a good indicator of the warmth obtained from all energy purchases. Changes in purchases of delivered energy, useful energy and expenditure on energy are examined to establish if there have been differential rates of change in the income groups. To a limited extent, these trends can be combined with the data on building fabric energy efficiency to quantify the total effect on low-income households. The links between domestic energy use and the emission of greenhouse gases are also explored.

Using economic theory, it is possible to infer whether low-income households value additional warmth by the extent to which they are prepared to spend extra income on fuel (Chapter 9). The effect of changes in the price of fuel or warmth can be similarly assessed and the crucial difference between these two commodities is confirmed. A model of household demand for warmth, utilizing the theory of consumer surplus, is extended and used to interpret research data. This provides an economic validation for a method of putting a monetary value on additional warmth — an area of considerable confusion in technical studies of the cost effectiveness of energy efficiency improvements. The way in which the value of warmth is included in the policies of different government departments is also examined, as an indicator of the consistency of resource allocation. Privatization of gas and electricity offers both constraints and opportunities.

The previous findings are brought together in Chapter 10 to define affordable warmth and, thus, provide a quantitative definition of fuel poverty. This also allows the relationship between the poor and the fuel poor to be stated. The policy initiatives that could comprise a scenario for the future are examined. These proposals identify the strategies needed and not the administrative detail of a programme for affordable warmth. There are changes in the political climate resulting from the privatization of electricity and gas and the creation of their respective regulators that affect the fuel poor.

Finally, an overview of the findings and the likely impact of existing policies is undertaken in Chapter 11 to summarize this complex field of empirical and policy analysis. There is a strong emphasis on the role of government in this book, because the regulatory and benefit structure is vital to the poor and directly influences the growth or reduction of fuel poverty. However, the increased awareness of the underlying causes of fuel poverty should enable many professionals, particularly those involved in the provision of housing, to contribute to the solution.

2 Emergence of fuel poverty

The present prominence of fuel poverty began in the 1970s after the increases in world oil prices in 1973-4. In order to assess the context in which fuel poverty emerged, levels of poverty, fuel availability and technical change in housing and heating systems are examined for the period prior to 1970. These areas are considered in further detail for the years 1970-5, together with evidence for the existence of fuel poverty. This sets the scene for an analysis of governmental concern since 1975: responses by government indicate both the seriousness with which policy-makers view a problem and their perception of the causes. This chapter, therefore, looks at the fuel poverty syndrome and political perceptions of it as an issue; the people that suffer are considered in Chapter 3.

Legacy of the pre-1970 years

Space rather than warmth

The housing stock at any one time represents a history of social expectations, building technology and fuel availability and provides a legacy of homes of varying suitability for current conditions. These factors are considered in relation to the housing stock that existed in 1970, and to common perspectives on living standards, and therefore warmth.

British housing and environmental health standards have not placed great emphasis on adequate heating. The primary concern in housing, ever since the first Public Health Act 1845, has been to avoid 'damp, structural instability, poor sanitation, fire risk and lack of light and ventilation' (Elder 1977, p.1). As a senior official at the DOE stated, 'Our traditional concern in this country has been to keep our homes dry, rather than warm' (Davies 1986). As many Victorian dwellings have fireplaces in most rooms, including bedrooms, there was no need for a legal requirement to provide a heating system: it was part of the structure. With the advent of alternative fuels to coal, the heating system is no longer an integral part of the building, so the risk of inadequate provision occurred. British legislation still does not require the provision of an adequate heating system in a domestic building.

Other European countries recognized the importance of warmth and designed their homes accordingly. The difference in standards was identified by Muthesius, a German writing in 1904 (but translated in 1979), when explaining why our homes are difficult to keep warm:

> the insubstantial structure of the English house, especially the meagre thickness of the walls, the absence of cellars, of double-glazed windows, the negligible care bestowed on ensuring that windows and doors fit snugly, the frequent absence of an entrance porch, the universal habit of using the whole attic floor for living accommodation. (Muthesius 1979, p.67)

There is little evidence that the British were aware of or concerned about these discrepancies and, as there were only local by-laws and no national building regulations prior to 1965, there was no mechanism whereby improved thermal insulation levels could be incorporated into building standards. The increasing construction of cavity walls at the beginning of the century appears to have been determined by builder and customer choice.

Legislation was passed in 1919 to enable local authorities to obtain subsidies when providing council houses (Dickens *et al.* 1985, p.148). For the first time, non-profit, social housing could be built. The emphasis was on space, light and fresh air:

> Council housing between the wars and in the 1940s concentrated on building suburban estates of cottage houses with gardens, some of which were located in pleasant settings and were built at generous internal and external space standards. (Dunleavy 1981, p.1)

The lack of adequate heating systems resulted in a misuse of resources in winter, which was a major concern of the government's Fuel and Power Advisory Committee in 1946:

> space in the home is of little value unless it can be used as living-space. In this country the principal faults of the past have been to neglect heat insulation in the construction of the house, and to limit space heating to one or two rooms... In cold weather the British home *is the smallest in the civilised world*. (Cmd 6762 1946, p.50)

Unfortunately, these comments, and the recommendations made by the Committee, did not result in improved levels of energy efficiency. Fifteen years later the Parker Morris Report was largely concerned with the same two issues: space and heating.

> During the greater part of the year, the two demands are scarcely separable, and it seems to us entirely wrong to go on building homes in which so much of the available space cannot be used for day-to-day activities throughout the year. We are quite sure of the widespread demand for a better heating service; even apart from the evidence it is shown by the figures for the purchases of various types of portable heaters. We therefore think it right to recommend that a considerably better heating installation should be adopted as a basic standard. (MOHLG 1961, p.3)

Because 'it is time to recognise that a home without good heating is a home built to the standards of a bygone age' (ibid, p.16), Parker Morris proposed, as a minimum, that the heating installation should be capable of maintaining 13°C in kitchens and circulation space and 18°C in living and dining areas; heating in bedrooms was optional and dependent upon finance availability. Though much applauded, the Parker Morris recommendations were not made compulsory for local authorities until nearly eight years later, from 1 January 1969 (MOHLG Circular 36/67), and then only to the minimum heating standard. The suggestion by Parker Morris that:

> Wherever family requirements will demand it, a more expensive installation capable of heating the bedrooms as well to 18°C (65°F) will represent the greater value for money. (MOHLG 1961, p.51)

was omitted from the mandatory standards covered by government subsidy (although, within the financial cost yardsticks, money could be used for this purpose). Thus, the large numbers of dwellings built in the 1960s did not have to comply with the standards in this circular and, even after 1969, the heating system in local authority dwellings had to be capable of achieving only the Parker Morris minimum. New housing, designed with a life of at least 60 years, was capable only of providing a minimal level of comfort. Cold, unused rooms help no-one.

Dual standards were operating with regard to indoor temperatures. After the particularly severe weather in early 1963, the need for adequate warmth at work had resulted in a clause in the Offices, Shops and Railway Premises Act 1963 requiring a minimum temperature of 16°C to be provided by employers. However, there was no requirement to provide a heating system in work areas in the home, such as kitchens, until 1969 and then the lower temperature of 13°C was stipulated for new local authority homes only.

Meanwhile, improvements to thermal insulation standards were slow. New housing in 1970 was only slightly better insulated than existing housing. Higher levels of insulation and improvements to the technical efficiency of heating appliances had been recommended. For instance, the Fuel and Power Advisory Council identified the need for more efficient appliances as well as 'insulation of the walls and roof' (Cmd 6762 1946, pp.2,34). The Parker Morris report included an appendix on double glazing, but did not recommend it because 'roof and wall insulation and weather-stripping of doors (and windows) are likely to give better value for money' (MOHLG 1961, p.73). However, the respective governments had not acted on the advice of their committees.

Fuel supply and choices

Solid fuel was the main domestic fuel for heating during all of this period. As coal was also the primary source of energy for the manufacture of both town gas and electricity, the availability of coal determined the amount of fuel used in the home for all purposes. Coal shortages, including the

rationing of domestic coal, continued for several years after the war, until
about 1950. With an increasing number of households, this represented a
declining level of use per household (Cmnd 8647 1952, p.20). In the early
1950s, solid fuel was still the recommended choice for domestic heating.

An alternative, and influential, solution was provided in 1953 when the
electricity supply industry introduced special tariffs for consumption
during off-peak hours at night and mid-afternoon, which encouraged the
development of space heating by storage methods (Electricity Council
1963, p.27). These tariffs could be promoted, despite shortages in peak
generating capacity, for instance during the harsh winters of 1961 and
1963.

Fuel substitution in existing houses occurred because of the Clean Air
Act 1956: to avoid the notorious and unhealthy 'smogs', an increasingly
large area of the country was zoned to use smokeless fuel, whether solid
or some other fuel. After solid fuel and electricity, town gas was the third
most important source of domestic energy and was used in at least 13m
homes by the mid-1960s, though not necessarily for heating. Conversion
to the cheaper North Sea gas started in 1967, but relatively slowly, so that
only 1.5 million homes had been taken off town gas by March 1970
(DUKES 1978, table 61). After the dominance of coal and fuel shortages in
the immediate post-war years, by the 1960s the householder was
confronted with a growing choice of fuels and systems in both new and
existing houses. A corollary of that wider choice was a greater range in
costs, though Wright criticized the advertising copy from the fuel
industries for implying that 'Everything is best' (1964, p.181).

An underlying perception, fostered by the electricity supply industry
since the mid-1920s, was that electricity would be a cheap fuel:

> The 'Weir Report' recommended the creation of the Central Electricity
> Board and of the national grid system. It was confidently predicted
> then, that by 1940, cheap electricity ... would be making other fuels
> obsolete. (Wright 1964, p.185)

The advent of nuclear power was expected to have a similar effect to the
installation of the national grid. The Chairman of the United States
Atomic Energy Commission stated in 1954 that:

> our children will enjoy in their homes electrical energy too cheap to
> meter. (Strauss 1954, p.9)

It is difficult now to estimate how prevalent this view was and how much
it contributed to both the acceptance of electric heating systems and to
expectations of cheaper fuel in the future, particularly as the opposite was
actually happening. Between 1962 and 1970, the real cost of electricity
and, to a greater extent, solid fuel increased, although gas prices declined
in real terms (DUKES 1974, p.127). However, it was recognized that living
standards rose substantially in the 1950s and 1960s, as summarized in
Harold Macmillan's 'Most of our people have never had it so good' in
1957. As a result, there was a general expectation of continuing
improvements in living standards, with cheap fuel a contributory factor.

Housing and heating provision

After the war, additional houses were needed partly because of the increasing numbers of households, as well as to replace war damage. House building methods were changing and local authorities constructed 500,000 non-traditional dwellings (see Glossary), mainly houses, between 1945-55 (AMA 1983, p.1). Because of the growth in industrialized methods and system building (both in Glossary), at least a further 750,000 local authority homes were built using these new methods, many of them flats in high rise blocks of more than five storeys (AMA 1984, p.24). Building these high-rise blocks was facilitated by the Housing Subsidies Act 1956, which encouraged multi-storeyed building (Murie 1983, p.60) and by the new forms of electric heating.

> High-flat building was the most extreme and conspicuous form of mass housing provision in this period, and has since become one of the most widely proclaimed (if unstudied) 'failures' of public policy in this field. (Dunleavy 1981, p.3)

A major reason for the failure was the mismatch between the type of heating system and the thermal insulation standards of the property. Because most of the new electric heating systems were based upon the efficient storage of heat, both within the appliance and the dwelling, the Electricity Council recommended that they should be installed only in dwellings with higher than average levels of insulation (Electricity Council, undated). Unfortunately, this advice was frequently ignored resulting in the installation of a heating system that was inappropriate for the dwelling (DEN 1 1977, p.6).

The Electricity Council's advice could be ignored because there was no standard or legal basis for determining the energy efficiency of the dwelling. Provided that the Building Regulations were complied with, there was no control over the relationship between the type of heating system and its suitability for particular buildings. This situation was exacerbated when the Ministry of Housing and Local Government (MOHLG) and later the DOE imposed a system of financial controls on public housing, called the housing cost yardstick, that required design changes at an extremely late stage, if tender prices were too high. In this situation, one easy solution for the architect was to reduce costs through replacing a heating system with high capital costs, such as a gas-fired one, with a cheaper system, usually electrically based. This appears to have occurred on numerous occasions and was criticized by a DOE working party, though the Department's own culpability was not mentioned:

> It must be accepted that electric heating should never be used in order to save capital costs. (ibid, p.6)

Despite uncertainty on actual numbers of dwellings affected, two basic statements can be made about the large number of new public sector houses built in the period 1945-70:

— before 1969 there was no minimum heating standard to be achieved;

— solid fuel was the main form of heating. Not until after 1964 was
gas heating chosen for a majority of new public sector dwellings
(Figure 2.1).

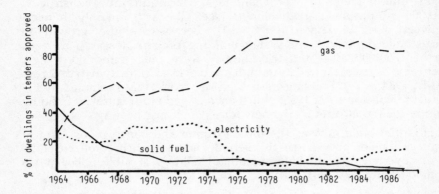

Figure 2.1 Space heating in new local authority dwellings, England and Wales
1964-87

Source: 1964-75 — DEN 1 1977, p.2; 1975-85 — DOE 1986a, p.88; 1986-87 — DOE
1989c, p.89

Progress on slum clearance and the rehabilitation of the existing
housing stock was also producing problems. The first national house
condition survey across all tenures of England and Wales took place in
1967 and showed that previous local estimates had seriously
underestimated the problem: 1.84m dwellings were classified as unfit in
1967, an increase of more than 1m over the figure of 0.82m in 1965 (Rollett
1972, p.303). The conditions in which the elderly lived were particularly
basic — 2m old people had access to an outside WC only (Shaw 1971,
p.14). No data are available about heating systems, presumably because
this was not a criterion of fitness. Thus, concern grew:

that the level of investment since the war had not succeeded in
extending benefits to the poorest, the worst housed and those with
least bargaining power. (Murie 1983, p.67)

The extent of poor housing conditions was being discovered, partly
through centralized, and therefore consistent, data collection.

Hypothermia

One of the extreme effects of living in cold conditions is that people can
suffer from hypothermia, when the body temperature drops dangerously
low — below 37°C. The incidence of hypothermia amongst new-born
babies was identified as 'a new syndrome' by Dr Trevor Mann (*Lancet*,
19 March 1955, letters). Further evidence accumulated of cases amongst

the very young and the elderly, particularly during the severe winter of early 1963, and of the extremely high mortality rate amongst hypothermic patients (*British Medical Journal*, 14 November 1964, p.1255). The link between poverty, cold homes and hypothermia was recognized:

> There is no doubt that in many cases old people do not have sufficient funds to spend on food and fuel, and during severe cold weather, as more money is required for fuel, there is less to be spent on food. (ibid, p.1258)

This may be the first authoritative reference to the 'heat or eat' problem — an indicator of fuel poverty. One reason for the new awareness was that:

> Most cases of hypothermia are missed because ordinary clinical thermometers do not record the condition. A standard thermometer which begins to register at 35°C (95°F) is misleading, as the actual temperature may be many degrees below. (ibid, p.1256)

This recognition resulted in more doctors testing for and finding hypothermia.

Rediscovery of poverty

> The belief in the early 1950s that poverty had been abolished had given way, over the next ten years, to a concern with a pocket of poverty among the elderly. (Bull 1971, p.13)

Research was undertaken to establish whether the problem was confined to the elderly, because:

> society has tended to make a rather sweeping interpretation of such evidence as there is about the reduction of poverty and the increase of equality since the war and, secondly, this evidence is a lot weaker than many social scientists have supposed. (Abel-Smith and Townsend 1965, p.12)

In particular, they challenged the work of Rowntree and Lavers that poverty had been virtually eliminated. Abel-Smith and Townsend found through careful and rigorous consideration of existing surveys that,

> the data we have presented contradicts the commonly held view that a trend towards greater equality has accompanied the trend towards greater affluence. (ibid, p.66)

One result of their work was additional public concern and the formation of advice and pressure groups such as the CPAG in 1965. These in turn generated a more informed debate and brought pressure on government, although initially they saw their role as that of primary helpers, not policy changers.

Income support for poor people was administered by the National Assistance Board after the Second World War. There is little available information on the payments they made, but 669,000 claimants received help with their heating costs in 1965 (Cmnd 6615 1976, p.20). By 1968, subsequent revisions had reduced to 143,000 those in receipt of HA, as the home was 'difficult to keep warm' or the claimant was housebound with a medical condition. These brief statistics indicate that heating problems were recognized and widespread, certainly by 1965.

Summary of conditions in 1970

In the last few years of the period to 1970, householders were getting a wider choice of heating systems and fuels than ever before: previously, solid fuel had been the main domestic fuel for heating, even in new buildings. As the level of thermal insulation required in the Building Regulations had not been materially increased, the majority of the housing stock in 1970 probably provided a similarly low level of energy efficiency. A unit of warmth cost almost the same for everyone. The most notable exceptions were the industrialized, high-rise flats with electric CH. Because of financial constraints on capital costs, a lack of bureaucratic safeguards combined with the electricity supply industry's competitive marketing, the level of insulation did not match the heating system provided and public sector tenants had housing that was predictably expensive to heat.

During the whole of the post-war building boom, there had been no standard for space heating in new local authority housing and even after Parker Morris became mandatory in 1969, the specification was basically for 'one warm room'. The delay in implementing the Parker Morris standards occurred despite the recognition that many households had difficulty with their heating costs and the payment of extra money for heating to over half a million claimants in 1965.

By 1970 it was known that there was widespread poverty in Britain and that for many, particularly elderly people, this meant living in cold conditions and being at risk of hypothermia. However, at this time, poverty and fuel poverty were not differentiated:

> To be able to keep warm without having to make inroads into one's budget for food and clothing would be, for very many people, a liberation. (Coates and Silburn 1970, pp.60-1).

1970-75: fuel poverty identified

Fuel price rises and inflation

In 1970 inflation went above 6 per cent per annum, after 15 years of fairly stable consumer prices (Lipsey 1979, pp.575 and 745). Under the statutory counter-inflation policy, introduced in November 1972 (*Britain 1975*, p.192), the fuel industries agreed to limit unit price increases to all consumers to 5 per cent per annum with a price freeze on standing charges. In the midst of these attempts to control inflation through subsidized prices, OPEC quadrupled the price of crude oil between 1973 and 1974, and the government announced a full return to economic pricing to be completed by April 1976 (Hansard 15 April 1975, col 295). The Labour Government's decision to permit energy prices to rise in line with costs was partly to limit public expenditure, but also in the interests of energy conservation (Chesshire *et al.* 1977, p.51). Thus, the demise of subsidies coincided with a period of rapidly rising fuel prices. This resulted in very substantial increases in fuel costs for domestic consumers: even with general inflation at 24 per cent in 1975, the fuel price index rose by 9 per cent in real terms. Public awareness of the price of fuel and its role in the domestic budget increased — energy costs had become headline news.

The annual change in the real price of gas and electricity from 1970 onwards is shown in Figure 2.2, using price indices. For the 20 years plotted, there were real increases in nine years for electricity and in five years for gas, demonstrating the benefits experienced by gas consumers. Over the entire period 1970-90 the fuel price index (FPI) increased 10 per cent more than the retail price index (RPI), making fuel relatively more expensive. Most of the increase has come from the rise in electricity

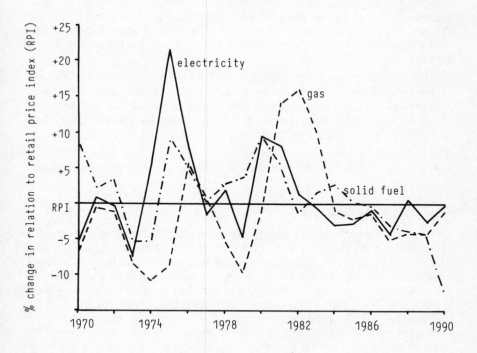

Figure 2.2 Annual changes in the real price of gas, electricity and solid fuel, UK 1970-90

Source: Based on price indices from DUKES

prices, which showed a real increase of 19 per cent , whereas the gas index dropped by 20 per cent in real terms.

With both fuels there have been extremely sudden price changes: a decline in real terms has preceded the greatest increases. Sharp price rises exacerbate budgeting difficulties for all consumers, particularly for low-income families, where fuel is already a large proportion of expenditure. In addition, social security benefit rates are fixed annually, so there can be many months between a fuel price rise and any resultant increase in financial support (or vice versa). For these and other reasons, rapid changes in fuel prices impact heavily on the poor. Injustice results if there are no mechanisms for smoothing out the necessary price increases, as Surrey and Chesshire (1979) argue:

> Graduation or 'smoothed' transition must therefore be a cardinal principle of energy pricing policy in order to comply with considerations of equity. (ibid, p.11)

The electricity price index represents a weighted average of changes to the two main domestic tariffs — off-peak and unrestricted. Off-peak sales represent 15-20 per cent of all domestic electricity consumption, so that the electricity price index can disguise considerable off-peak price increases. For instance, the off-peak tariff went up from 0.54 p/kWh in August 1974 to 0.90 p/kWh in April 1975 (Electricity Council, pers comm) — a 67 per cent rise in eight months and a 50 per cent increase in real terms.

> The resultant outcry, culminating in a parliamentary debate, persuaded the government to subsidise part of the extra fuel cost to off-peak electricity customers, in order to maintain the previous percentage differentials between full price and off-peak electricity for a further period. (NCC 1976, p.4)

The promise of new technology using cheap electricity had, in reality, become an inappropriately installed heating system using an increasingly expensive fuel.

Changing to credit metering

One further result of the changing domestic fuel mix was a general change from prepayment for solid fuel, to credit metering for gas and electricity. In addition, in order to reduce collection costs, British Gas and, to a lesser extent, the Electricity Boards, phased out many prepayment meters, so that between 1966-76 the number of customers that were transferred from pay-as-you-go to three-monthly credit meters was 3.9 million (NCC 1976, p.63; Hansard, 28 November 1975, WA col 330).

Whilst this change in billing procedures does not adversely affect the price of fuel — the cost is less with credit meters — it creates substantial problems for low-income households and the ease with which they budget for fuel, for two reasons. First, the amount of money to be saved up in thirteen weeks is equivalent to at least two weeks' total income (based on the *Family Expenditure Survey* — FES), which is a considerable challenge to a family on a small weekly or two-weekly benefit. Secondly, experience based on the amount of usage purchased by a particular coin (in a prepayment meter) is not easily transferred to the cost of a number of therms or kilowatt hours with a credit meter designed only as a flow gauge: only the quarterly bill gives the unit cost.

For people on a low income it is particularly important to keep debts small, otherwise repayment becomes an almost impossible burden. With such infrequent billing and with primitive metering methods, households can easily slip into unintended debt, and disconnection often occurs when consumers accrue debts they cannot repay. The Select Committee on Nationalised Industries considered that:

> the interests of the gas and electricity industries have counted for more than those of the customers. (HC 353 1976, p.lxxxv)

and that hardship and debt were resulting from the industries' insistence on moving to credit meters.

Heating additions

By 1971, about 7 per cent of all supplementary benefit (SB) claimants received HA (Table 2.1), and over 80 per cent of heating payments went to pensioners on SB (Cmnd 6615 1976, p.21). The allowances were paid at three different rates, based on a rough equivalence to the cost of 0.25, 0.5 and 0.75 cwt (hundredweight) of coal per week (*Cold homes: the crisis* 1982, p.22). Thus, even at this stage, the DHSS recognized variations in the cost of heating.

Further amendments took place in 1973, prior to the oil crisis, 'as a result of continuing concern about heating and the risk to older people of hypothermia' (Cmnd 6615 1976, p.22). By 1975, claimants with CH were entitled to an exceptional HA, reflecting problems with electric systems. Between 1971-5, the number of claimants in receipt of SB was fairly stable, so that the increase in HA resulted from changes in eligibility. The number had more than quadrupled since 1971 and HA were given to a third of all claimants by 1975, 80 per cent going to pensioners. The further rise and fall of heating additions is discussed below. However, the cost to government of fuel poverty was rising.

Disconnections

Despite the growth in the payment of HA, disconnections were also rising. In 1975 a record total of 176,000 households lost the use of electricity or gas (Figure 2.3), apparently as a result of the real fuel price rises and increasing credit. The most obvious explanation, though not necessarily the only one, is that HA went mainly to pensioners on SB, whereas few of the elderly get into debt and are disconnected. Therefore, HA could help the pensioners without having any influence on the numbers with fuel debts. The relationship between real price movements for gas and electricity and disconnection figures can be seen by a

Table 2.1 Numbers in receipt of heating additions and total cost, GB 1970-87†

	Total SB Claimants (000)	Claimants in receipt of HA Number (000)	%	Cost of HA (£m)
1970	2740	150‡	5	-
1971	2910	194	7	-
1972	2910	232	8	-
1973	2680	503	19	11
1974	2680	708	26	20
1975	2790	915	33	36
1976	2940	1233	42	61
1977	2990	1455	49	83
1978	2930	1545	53	90*
1979	2850	1638	57	108*
1980	3120	2037	65	161*
1981	3720	2347	63	264*
1982	4270	2551	60	325*
1983	4349	2587	59	380*
1984	4609	2733	59	400*
1985	4700**	2829*	60	430*
1986	4938	2799	57	508
1987	4896	2710	55	494

Notes:
† Heating additions were abolished from 1 April 1988
‡ Estimated from Cmnd 6615 1976, p.21
* DHSS pers comm 11 Feb. 1988, costs for financial years, e.g. £90m 1978/9
** Author's estimate

Sources: 1970-87 — SSS Table 34.44

comparison between Figures 2.2 and 2.3. The causes of variations in disconnection rates are certainly complex and are discussed further below, but the 1975 peak was believed by many to be causally related to the fuel price increases. However, total disconnections were at a rate of nearly 150,000 a year from 1970 — before the first oil crisis.

The rise in disconnected households in 1975 focused public attention on the ability of a publicly-owned monopoly supplier of an essential commodity to disconnect debtors without a court order.

No other public supplier of essentials withdraws supplies from a whole family because the head of household is in debt ... the water industry has the powers ... but ... debts are pursued through the courts. (NCC 1976, p.81)

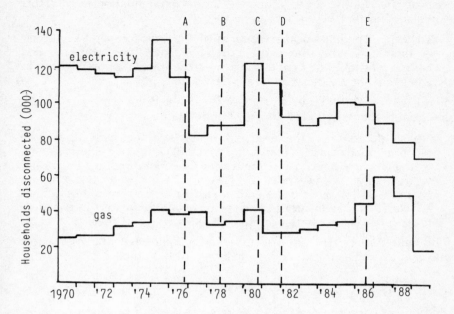

Figure 2.3 Households disconnected from electricity (England and Wales) and gas (GB) 1970-89

Notes:
A: Code of Practice introduced; B: 1st revision; C: 2nd revision; D: 3rd revision; E: British Gas privatized

Source: 1969-72 — Hesketh 1978, pp.13, 133; 1973-75 — NCC 1976, pp.74-5; 1976-82 — NRFC 1982; 1983-89 — Electricity Consumers' Council, Gas Consumers' Council quarterly reports

Cold homes

Other evidence of severe hardship continued. To investigate the incidence of hypothermia amongst the elderly, a detailed national study of 1,000 old people was initiated by David Donnison, Director of the Centre for Environmental Studies (Wicks 1978, p.xii). The survey, undertaken in 1972, found that many of Britain's elderly lived in cold conditions and 9.6 per cent of the sample were at risk of developing hypothermia because their body temperature was so low (ibid, p.158). Hypothermia is the extreme result of living in cold conditions and during the late 1960s and early 1970s there was increasing awareness that many other diseases and causes of death are cold-related (Chapter 7). In order to minimize the number of unnecessary and avoidable winter deaths, the DHSS issued guidelines for use by volunteers and social workers when visiting the elderly. These reiterated the recommendation in MOHLG Circular 82/69 that:

> To keep old people warm in winter the living room temperature should be about 21°C when the temperature outside is −1°C. Bathrooms and bedrooms should be kept at the same temperature if possible, but in any event should be kept warm. (DHSS 1972, p.3)

In the winter of 1974, a survey for CPAG examined the heating resources and requirements of 18 low-income London homes:

> No single room in the nine old people's homes was up to the recommended standard of 21°C... Two families and one elderly person were living in homes where the warmest temperature on a fairly mild winter's day (9°C) was only 10°, 11° and 12°C respectively... Only 16 rooms in all 18 homes had an adequate heater, and six homes did not have an adequate heater in any room... Half the heaters in use were those most expensive to run, namely electric fires. (CPAG 1974, p.11)

This limited survey provided documented evidence that families on a low income, as well as the elderly, were living in cold homes.

Public pressure

Because of concern over the social repercussions of the fuel-suppliers' policies on disconnections and in response to the increasing number of fuel related problems being presented to advice agencies, a presssure group was formed in October 1975 — the National Right to Fuel Campaign (NRFC). Membership grew rapidly to include nearly 200 local fuel groups and 15 national organizations such as CPAG and Age Concern (NRFC 1977, p.1). Fuel poverty had become a recognized problem. During the fairly cold winter of 1975-6, the

> fuel problems of people of all ages were debated on TV, radio, and in newspapers. Hypothermia provided the front page story in two popular tabloids on the same day. (Gray *et al*. 1977, p.1)

In addition, advice agencies were reporting a substantial increase in fuel debt problems and found abundant evidence of hardship to individual households, 'many of whom will have not previously had difficulty in paying their quarterly accounts', though 'fuel debts are often associated with other debts, such as rent arrears' (NCC 1976, pp.10-12).

Summary of developments in 1970-75

The three-day week and reduced electricity supply caused rising public awareness of energy issues, which was reinforced by the OPEC crisis and the rapid increase in fuel prices that followed. The response to high disconnection rates and surveys of cold homes was to see them as a result of high fuel prices, and by implication, as a new phenomenon. However, as the evidence provided above demonstrates:

There is in fact little evidence that the increase in prices for fuel in 1974-75 led to a *new* problem of fuel poverty. (Bradshaw 1980, p.22)

Fuel poverty was the new name for an old problem: comparable numbers of people had suffered from hypothermia and had been disconnected in the past, and HA were paid as part of National Assistance at least since 1965. The rise in fuel prices, after October 1973, was an aggravation, but not the root cause of fuel poverty. Thus, the perception of the causes of fuel poverty and the real reasons differ, though Marigold Johnson, in one of the first references to fuel poverty, was close to the truth when she stated that fuel poverty is caused by 'society's failure to plan for an age of high-cost fuels' (DEn 1976b, p.34)

The low levels of energy efficiency in the housing stock meant that warmth was an expensive commodity in Britain. Real fuel price increases make warmth even more expensive and the greatest rises were in coal and electricity prices — the two fuels most likely to be used by low-income households for heating. Had there been a policy for warmth in the past, leading to a higher level of energy efficiency throughout the housing stock, fuel price increases would have had less impact. (This interrelationship is summarized in COWI, and examined in Chapter 5.) Government inaction on warmth-related housing standards cannot be explained easily, but government advisers had been advocating improved insulation and efficient appliances since at least 1946.

Instead of lowering the cost of warmth, some changes in heating systems, particularly electric CH, had had the opposite effect. The high cost of keeping warm that resulted from this combination of factors would have caused difficulties for households on even quite substantial incomes:

> Much of this poverty Britain has created for itself. We have too often consigned poor and vulnerable people to high-cost environments. There richer people could manage very well, but the poor, who cannot afford it, find that they *have* to have central heating — the most expensive forms of it moreover, because all the flats have electric night storage (or underfloor or ceiling) heaters, and there are no open fires. (Donnison 1982, p.3)

As these acute problems occurred in local authority housing and the tenants were on relatively low incomes, fuel poverty became seen as a problem of shortage of income, rather than as a problem of high cost warmth. But, by the mid-1970s, fuel poverty was an acknowledged social issue.

1976-90: government responds

The seriousness of the situation can be gauged by the publication, in 1976, of three government-sponsored reports on aspects of paying for fuel by poor consumers, confirming that fuel poverty was seen as a fuel-pricing problem. The Department of Energy (DEn) was unable to propose any 'satisfactory' way in which the tariff structure could be altered to help poor consumers and thus made no positive recommendations (DEn 1976a). The two other reports recommended that the industries should not be able to disconnect domestic debtors and that money should be recovered by other means, such as installing prepayment meters (DEn 1976c, para 167; NCC 1976, p.82). The fuel boards reacted strongly, claiming that the increase in debt burden and court costs would cause price rises of at least 10 per cent, further fuelling inflation (Cooper 1981, p.70). Not surprisingly, no legislation was passed, although external best estimates of the impact were that prices would rise by about 1 per cent (NCC 1976, p.84). However, British Gas and the Electricity Boards did agree to a voluntary Code of Practice, effective from January 1977, to minimize hardship and clarify consumers' rights.

Three years and three reports after the 1973 energy crisis, the voluntary Code was the only tangible achievement of energy policy aimed at helping the fuel poor. The Labour Government had encouraged the debate about billing and pricing procedures and their effect on the poor, through the publication of these reports, but then failed either to act or to provide alternative explanations of the causes of fuel poverty.

Since 1976, the government has not published any reports, either from individual departments or interdepartmental in character, on any aspect of fuel poverty to establish or even to discuss the priorities for policy. A study undertaken in 1985/6 'was prepared only as a basis for policy advice for Ministers' and thus not published (HC 262 1986, p.xviii). The Select Committee for Energy's request that it should be placed in the House of Commons' library (ibid, p.vi) does not appear to have been heeded.

Rise and fall of disconnections

The link between disconnection rates and real fuel price rises was originally seen as a major indicator of fuel poverty. However, the evidence before 1974 and recently shows that the issue is more complex. Disconnection rates can rise when the price of the fuel is falling (gas 1980) or vice versa (gas 1982). Similarly, gas disconnections can increase as electricity's are declining (1987) or the other way round (1978). A further confounding factor is that the supply industries' disconnection rates and policies may be inversely related to public concern: because of external factors, disconnections become the concern of senior management, so there is a policy shift. This would, for instance, explain the drop in disconnections during discussions on the introduction of the Code in 1977, or the doubling of British Gas disconnections over the period immediately before and after privatization, when the primary focus was the balance sheet (Figure 2.3). This doubling resulted in considerable concern and brought disconnection to the attention of the Director General of Gas Supply (the Regulator). As a result, the statutory Licence now includes a condition on methods for dealing with customers in default and a parallel requirement is imposed on the privatized electricity industry. By 1989, disconnections in both industries were at their lowest recorded level, though still at a combined total of 90,000 a year. Despite the uncertainty about the effect of different causes, changes in real fuel prices are only one of the factors.

Fuel-pricing policies

Originally, the main conflict between fuel poverty campaigners, and the DEn and the fuel industries was whether fuel-pricing policies should recognize social constraints or be based solely on economic criteria. The complexity of the issues involved, combined with the lack of data from the supply industries, has made it difficult for fuel poverty policy analysts and campaigners to criticize constructively, particularly when most of the debate has originated with social welfare specialists. Support came, briefly, when Tony Benn was Secretary of State for Energy in the 1974 Labour administration and suggested:

> energy policy should ensure that every one can afford adequate heat and light at home. (DEn 1976a, p.1)

Though his statement was contradicted, in the same document, with the traditional DEn view:

> Protecting the poorer members of society and ensuring that they are in a position to buy adequate warmth is not a direct element in *energy* policy as such; it is essentially a problem for the social services and for other policies affecting the distribution of income. (ibid, p.22)

A Fabian pamphlet written by economists and energy advisers did consider energy policy in relation to fuel poverty (Goode *et al.* 1980). Their conclusion, that fuel-pricing policy should not be used as a mechanism for fuel poverty assistance, was made by energy specialists sympathetic to the injustice of fuel poverty. The importance of investment in the efficient use of energy, with industry funds, as suggested by Goode, is again being recognized in the 1990s.

Since 1979, fuel prices have risen as a result of non-energy related objectives (MacKerron 1987, p.5) on three occasions:

— in January 1980 the government required gas prices to rise by 10 per cent over the rate of inflation for each of the next three years;
— political intervention also resulted in electricity prices rising by 2 per cent in 1984, when the electricity supply industry wanted to keep them stable (HC 276 1984, p.ix);
— the Secretary of State announced in December 1987 that electricity prices would rise by around 15 per cent over the next 15 months, allegedly to raise capital for new power station construction, prior to privatization.

On none of these occasions has the non-energy related price increase been of benefit to disadvantaged consumers, though social policy objectives could be met in this way:

> Insofar as they place a comparatively heavy burden on those who are relatively poor, prices that create large financial surpluses conflict with

the widely accepted principle of progressive taxation designed to promote more equitable income distribution. On this, as on other pricing questions, there is a choice in the extent to which the consumer or the tax-payer should pay. (Surrey and Chesshire 1979, p.10)

Any pricing procedure has to acknowledge that this process occurs and that generating surpluses, whether for the Treasury or shareholders, impacts on the fuel poor. The opportunites for government intervention post-privatization are curtailed as both electricity and gas prices are determined by formulae, as part of the powers of the regulators (Ofgas and Offer). The main route for government to affect fuel prices must now be through taxes (eg VAT or carbon taxes). The effect of the regulation formulae has yet to be deciphered, particularly as they vary for the fourteen public electricity suppliers (PES — old area electricity boards). Theoretically, fuel-pricing could still be used to assist low-income households, but it is assumed that there are too many other more powerful objectives. Therefore, fuel-pricing policies are not a likely component of anti-fuel poverty policies; the more likely problem is that taxes on fuel will increase fuel poverty.

Additional income

Ostensibly to offset the hardship caused by the cessation of subsidies, though perhaps to appease public opinion, in 1977 the government belatedly introduced an electricity discount scheme for low-income domestic consumers. This was the first specific fuel benefit to come out of the DEn's budget and continued for three years. Meanwhile, the DHSS were paying increasing amounts in HA as eligibility was extended. From 1978, over half of all SB claimants received a HA, although they were meant to be paid in exceptional circumstances only (Table 2.1). Difficulties with heating costs were no longer the exception.

Until April 1988, HA were provided, if eligibility could be established, for one of the following reasons:

— a member of the claimant's family was retired, young, sick or disabled;
— the property was centrally heated (no new applications after August 1985);
— the property was difficult to heat (e.g. it was damp, or large);
— the fuel used in the central heating system was particularly expensive (mainly preserved, historic, off-peak electricity tariffs).

About 2 per cent of HA were given for the last two categories, and approximately 40 per cent because of the presence of CH, so that the majority (58 per cent) of extra payments were paid on the basis of the social characteristics of the household. This latter form of eligibility was administratively easy but only gave the greatest assistance to those in the greatest need, if there was a clear link between the social characteristics of the occupants and high heating costs from energy inefficient homes: a relationship examined in Chapters 4 and 5. Even if a correlation does appear to exist, as between elderly people and cold conditions, providing extra income may not be cost effective in producing warmer homes.

The provision of HA, however, gave real help to some of the poorest families in the country. The total amount of money paid annually rose from £11m in 1973 to £494m in 1987: a ten fold increase in real terms (Table 2.1). Half of the growth was due to the greater number of claimants in receipt of an individual HA, often because of economic or demographic changes. The other major influence was the additional eligibility required by continuing evidence of hardship (Boardman 1986a, p.70). By 1987 the government had paid a total of £3.4bn in HA — all of it being used to mitigate the immediate effects of fuel poverty, none going on preventative investment.

From 1 April 1988, SB became income support (IS) and all HA were abolished. Those that were based on the social characteristics of claimants (mainly age) were, in theory, incorporated in the new premiums. With the numerous changes taking place, it has proved impossible to identify what actually happened to this £300m. The remaining £200m was being paid largely on the basis of the heating system and fuel used and no attempt was made to replace them in the new scheme, so the money was lost to claimants. These were abolished, according to the Secretary of State for Social Services, because modern CH systems are more efficient than old ones used to be and because people with CH spend less than those with non-CH (Cmnd. 9518 1985, p.21). There has been a negligible improvement in the efficiency of CH systems (Henderson and Shorrock 1989, pp.20-1) and CH users do spend more than consumers with non-CH (Evans and Herring 1989, p.87). Thus, both the reasons given for the abolition of energy-efficiency related HA were fallacious.

The implications for energy efficiency policy of the payment of HA were recognized by the Rayner Scrutiny:

> In issuing heating additions ... the Government in effect partly acts as an energy user, in that it helps meet energy bills. But it does very little through energy efficiency measures to reduce those bills (or to allow improved levels of comfort for the same expenditure on energy — many recipients of heating additions underheat their homes considerably). In some cases the very reason the heating additions are needed is that the energy is used highly inefficiently because of poor heating systems or insulation standards. (Finer 1982, p.38)

However, HA have been abolished without an increase in capital investment, thus compounding the injustice.

Energy efficiency policy

As explained in Chapter 1, the concepts of energy conservation and energy efficiency have been gradually evolving, particularly since 1973, though there was a delay in formulating effective policy. In a major statement on energy saving initiatives, the measures announced by the Secretary of State for Energy involved 'urgent discussions' with local authorities and doubling thermal insulation standards in new dwellings (Hansard 9 December 1974, cols 27-31). For fuel poverty sufferers these policies were of no immediate help. No further policies to help consumers save energy in their homes were announced for another three years:

> This was due partly to the sheer lack of information both in government and elsewhere about energy-using activities and the potential or options for energy conservation. (Bending and Eden 1984, p.242)

The package announced by the Labour Government in December 1977 represented the first real evidence of a major commitment to a national energy conservation programme (ibid, p.243), and then the emphasis was on energy saving, by all income groups, rather than on providing specific assistance to poor families. This was the first and last major initiative from the DOE.

A further reason for the lack of emphasis on energy demand and conservation was Britain's substantial energy resource endowment, including North Sea oil and gas, which encouraged:

> complacency borne out of relative plenty (the UK as 'an island of coal set in a sea of oil') and effective self-sufficiency in energy. (Chesshire 1986, p.403)

This was supported by the 'traditional fixation of the energy institutions and policymakers in the UK with the *supply* side', together with the lack of a coherent energy strategy and almost sole emphasis on the price mechanism (ibid, pp.403-5). The Conservative Government's view was summarized by Nigel Lawson, when he was Secretary of State for Energy:

> Conservation ... is a way for the consumer to cut his costs. It is unlikely in the extreme that we would be better off if decisions about insulating millions of homes ... were all made within Whitehall. The Government's role is neither to induce the individual to take decisions against his better judgement, nor to waste public money in subsidising investment that is already well worthwhile. (DEn 1982a, p.6)

The assumption is that all consumers have the financial ability and legal rights to respond by undertaking energy efficiency improvements to their home — a situation that does not exist for low-income households or people in rented accommodation.

Part of the problem with policy towards the efficient use of energy in the domestic sector has been, and still is, the confusion over departmental responsibilites. This was highlighted when the EEO was created in November 1983:

> The Secretary of State for Energy will take on overall responsibility for promoting energy efficiency in buildings ...
> The Department of the Environment ... will continue ... to be responsible for the energy efficiency of domestic buildings generally.
> The Secretary of State for Health and Social Security and the Secretaries of State for Energy and the Environment will continue to co-ordinate

policy on energy efficiency with particular relevance to the disadvantaged. (DEn 1983a, pp.8-9)

These ambiguous definitions indicate the uncertain interface between the DEn and the DOE, particularly over policy for the energy efficiency of heating systems and whether fuel poverty policy is part of housing, social or energy policy. It would have assisted the development of discussions and debate if these responsibilities could have been clarified and stated more clearly. The important difference between the energy efficiency *of* buildings and energy efficiency *in* buildings was not clarified then, and has not been apparent in subsequent policies. It is still not certain whether any department is concerned about the economically efficient use of energy in domestic heating systems.

> Too often poor people were caught in the cross-fire of frontier warfare between bureaucracies, and ended up getting the wrong help or none at all. (Donnison 1982, p.41)

Reports of hardship

Case studies of households with fuel debts or those disconnected have continued to find severe problems (Lorant 1981; Winfield 1982; National Council for One Parent Families 1978 and 1985). Though often small samples, these reports catalogue the human cost of disconnection and the awful suffering caused to all members of the family, especially the children, when the only source of heating is human warmth:

> They sit on my lap, inside my coat, and go fast to sleep. The boy goes on my lap and the girl goes on hers. They keep their bodies close up to us to get warm, then they go to sleep. (Winfield 1982, p.21)
> We all three slept in one single bed — it was warmer than letting them sleep on their own. (ibid, p.24)
> We all had to go to bed at half past five. (ibid, p.33)

Between 1975 and 1983, the National Association of Citizens Advice Bureaux conducted five national surveys on fuel debts and fuel-related problems. The 1983 report found that 'although fuel poverty continues to be a major concern to the CAB Service, the Code of Practice is almost irrelevant for many bureaux handling fuel problems' (NACAB 1983, p.3) and 'the suffering caused by inadequate fuel supplies (to the individual household) is as widespread as ever' (ibid, p.56). The focus on debts, disconnections and the amount of fuel expenditure probably reflects both the evidence presented to advice agencies and the ease with which the data can be collected. Surveys of actual heating levels are expensive and complex to mount, so there has been no sequel to the detailed survey of the old and cold undertaken in 1972 (Wicks 1978), though the Institute of Gerontology are undertaking a new survey in 1990.

The problem of fuel poverty, particularly for the elderly, is regularly debated in the House of Commons, as well as covered by the reports of Select Committees (usually Energy). Despite this continuing evidence of both hardship and concern, there has been no further consideration of fuel poverty in published government reports, not even during 1986 — Energy Efficiency Year. Instead, the Secretary of State for Energy and the Minister for Social Security have both denied that fuel poverty is distinct from general poverty (quoted in Chapter 1) and HA have been abolished.

The political rationale behind the review of the benefit system and the considerable level of hardship resulting from the changes have been well catalogued (Andrews and Jacobs 1990). Not only have many people come off benefit and the rates for most suffered a real decline, but many changes are to be phased in during a transitional period. The government's undertaking that no claimant will lose money in absolute terms, results in them suffering real cuts, as present benefit payments are frozen. For many households, state-provided income will not increase in real terms for several years.

This continues the trend identified for 1979-87. Including housing costs, the real incomes of the bottom two deciles of the population decreased by 1.1 per cent, whereas the rest of the population had a 29 per cent increase (Government Statistical Service 1990, p.52).

Conclusions

British housing standards have traditionally placed more emphasis on space and freedom from damp than warmth. Therefore, insulation has been required in the building fabric only since 1974 and a set heating standard for new local authority housing existed only for the years 1969-80. If there is a link between fuel poverty and the energy inefficiency of the housing stock, then the regulations governing housing standards and environmental health have allowed the preconditions for fuel poverty to develop in Britain.

The incidence of hypothermia and disconnection levels prior to 1973-4 indicates that fuel poverty existed before the first oil crisis and official concern had resulted in help with heating costs going to over half a million households in 1965. The hardship caused by the real fuel price increases of the mid-1970s created the impression that they were the primary reason for fuel poverty. Undoubtedly higher energy costs make fuel poverty worse and the 10 per cent real increase in fuel prices over the period 1970-90 has contributed to the suffering.

The third element in the fuel poverty equation is income levels and most of the poor have suffered real losses in income, certainly since 1979 and as a result of the 1988 social security changes. Thus, all three of the main suspects for causing fuel poverty — housing conditions, fuel prices, income levels — have deteriorated since 1970 and provided the basis for a growth in fuel poverty.

One reason for government inaction may be a genuine uncertainty about the most appropriate policies to assist fuel poverty sufferers, though the lack of published, official studies, since 1976, indicates that fuel poverty is not an area of active concern. Whether cause or effect, there is no coherent policy responsibility in this area for either the DEn or

DOE, despite the creation of the EEO in 1983. Largely because of the lack of action in other government departments, most of the assistance received by fuel poverty sufferers came from the DHSS in the form of HA. With the abolition of HA in April 1988, the DSS is, in effect, stating that it will provide income support based on social characteristics, but not variations in energy efficiency. There is no evidence that the DOE or DEn are proposing adequate, compensatory policies. Thus, the fuel poor appear likely to be left in a policy vacuum — adequately warm homes are not the responsibility of energy, housing or social policy.

3 Profile of the fuel poor

To establish whether there is any difference between poverty and fuel poverty involves two separate questions: first, are the groups that suffer the same or not and, secondly, are similar remedies required? This chapter is concerned with the first question and focuses on the people involved. Most of the data, however, are for households or the head of household. Partly this is a statistical necessity, but, more importantly, it is appropriate to consider people in the groups that occupy a house and use the same heating system. Before considering the families in poverty, the numbers and characteristics of those with fuel bill problems are examined. Later chapters consider the second question — the possible causes of fuel poverty and thus the policy responses needed.

The two government ministers quoted in Chapter 1 maintained that fuel poverty is just one aspect of general poverty and both can be alleviated through the social security system. However, Chapter 2 has demonstrated that perceptions of the causes of fuel poverty (such as large fuel price increases) are not supported by the evidence — the extent of hypothermia and levels of disconnections have not been affected by real fuel price rises or falls. Therefore, existing perceptions cannot be relied upon and the relationship between poverty and fuel poverty needs to be analysed from first principles.

The analysis is hampered by the lack of a quantifiable definition — all that can be stated is that fuel poverty sufferers have either cold homes, or large fuel bills (relative to income), or both. Expenditure data on its own cannot provide any information on the temperature achieved in the home. For instance, low levels of energy consumption would provide adequate warmth in a well insulated home, but not in a poorly insulated one. Therefore, this chapter can provide only partial information on the characteristics and numbers of people known to suffer from fuel poverty, because they have identifiable fuel bill problems. Before April 1988, the receipt of HA provided a wider definition, as it affected 2.7m households in 1987 (Table 2.1). However, this was never a very reliable guide, as 60 per cent of HA were given on the basis of the claimant's social characteristics.

Despite the difficulties experienced in identifying the fuel poor as a group, a working definition of fuel poverty has to be established, in order to focus the analysis on an appropriate section of the population. Because the poor, by definition, have difficulty in purchasing sufficient necessities, including fuel, for a modern standard of living, the initial assumption is that all the poor are fuel poor to some extent. Therefore, some of the ways of defining poverty are considered to establish which is the most applicable and, provisionally:

the number that suffer and therefore the size of the problem;
whether policies could be targeted through the benefit system.

In order to answer these questions, the relationship between poverty, fuel poverty and receipt of benefits has to be unravelled. The identification problem is demonstrated by the two illustrative Venn diagrams (Figure 3.1): at the moment, neither the relative sizes of the groups nor the extent of overlap are known. The answer has to wait until Chapter 10, when a basis for quantifying fuel poverty is established.

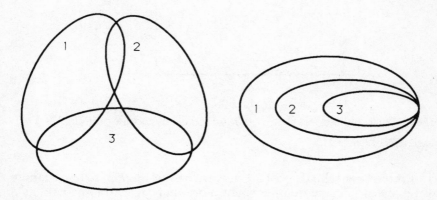

Figure 3.1 Venn diagrams depicting two possible relationships between the families in poverty, in fuel poverty and on benefit

Notes: 1 — in poverty
            2 — on income support or housing benefit
            3 — in fuel poverty

Problems with fuel bills

Disproportionately large expenditure

The proportion of total expenditure that goes on fuel varied from 11.5 per cent for the poorest quintile to 3.5 per cent for the richest in 1988 (Figure 3.2). Even amongst low-income households, the proportion spent on fuel varies considerably, with pensioners (14.4 per cent) and single parents (15.8 per cent) allocating the largest budget share to fuel in 1987 (Social Trends 1990, p.99). More details on fuel expenditure are given below and in Chapter 8, when the definition of poor used in this study has been established.

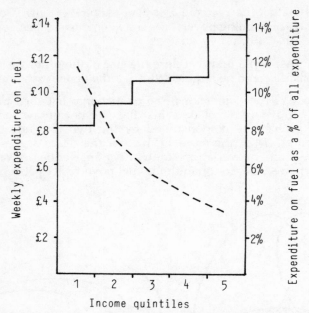

Figure 3.2 Household weekly expenditure on fuel by income quintile, UK 1988

Source: FES 1988, p.21

The average household spent 5.1 per cent on fuel in 1988, so that a family in the lowest income quintile spent more than twice the median, as a proportion of income. In an analysis of the 1977 FES, two DHSS economists took twice the median as the point at which disproportionate expenditure begins to occur with the basic necessities of food, housing and fuel (Isherwood and Hancock 1979, p.5), and, by implication, undue financial hardship. They found that 'virtually no households have a proportionate expenditure on food greater than twice the median, whereas about 18 per cent of households spend disproportionate amounts on fuel and housing' (ibid, p.8). Some characteristics of the 18 per cent of households with disproportionate fuel expenditure are shown in Table 3.1. Isherwood and Hancock's work is based on the FES, which is widely accepted as being a representative sample, so the results can be grossed up to the whole population. Changes in the intervening thirteen years cannot be estimated. The largest category is households who have qualified for a rent or rate rebate (the predecessor to HB), because their income is assessed as being low relative to their housing costs. This is particularly revealing as the households are distributed across all income quintiles. No further evidence is provided, because none has been identified, on whether the same households have relatively high housing and fuel expenditure, but the flexible matching of needs and means that occurs with HB is apparently necessary for fuel costs as well.

Table 3.1 Characteristics of head of household with fuel bill problems (%)

	Great Britain Disconnected 1980-81	Debtor* 1980-81	On fuel direct** 1982	UK Households with disproportionate fuel expenditure 1977
ECONOMIC STATUS				
With children	73	77	75	
Worker	42	53		36
Adult at home				62
Pensioners	2	5	7	49
On SB	32	28	100	
On HB or equivalent				86
Income in Q1***				42
Q2				26
Q3				13
Q4				11
Q5				7
HOUSING CONDITION				
Renting from LA	72	65		40
All renting				55
CH	59	59		43
NATIONAL TOTAL (000)	152		189	3,600

Notes: Only the five income quintiles are additive. The national totals refer to data from Table 3.2, not the numbers in the samples.
* As defined by Berthoud 1981, excludes disconnected households.
** When a deduction is made from SB at source to cover both fuel arrears and current consumption.
*** Gross weekly income in the lowest quintile.

Sources: Cols 1 and 2: Berthoud 1981, pp.127, 202-3, 209-10, 214; Col 3: Berthoud 1984, p.A18a; Col 4: Isherwood and Hancock 1979, p.20

Another group with disproportionately high fuel bills are households headed by 'an adult at home', for instance, because they are sick, disabled, retired or unemployed (Isherwood and Hancock 1979, p.19). This characteristic is a good indicator of relatively high fuel bills as only 22 per cent of the households with fuel expenditure below twice the median were headed by an adult at home — a threefold variation. The other columns in Table 3.1 are discussed below.

Fuel debts

It is not easy to define when someone is in default with a fuel bill. The majority of gas and electricity customers have credit meters, are billed in arrears and could be considered in debt if they do not pay up quickly. Berthoud introduced the useful distinction between 'late payers' and debtors (1981, p.185). This effectively distinguishes between those that are manipulating the system in order to obtain further credit and those that are in genuine difficulty. According to Berthoud, the likelihood of being a debtor (whether disconnected or not) is closely associated with low income, whereas late payers were found in all income bands (ibid, p.211). Electricity and gas debtors are at risk of being disconnected, and because this is such a powerful sanction, it can be assumed that a continuing fuel debt is evidence of an inability to pay.

The number of households being disconnected fell below 100,000 in 1989 for the first time since at least 1970 (Figure 2.3 and Table 3.2), but these represented a relatively small proportion of the households with fuel debts. To avoid disconnection, some households are provided with a prepayment meter, often calibrated to recoup an outstanding debt. Since 1977, over 2.5m prepayment meters have been installed. Originally these would have been coin operated, but increasingly coinless prepayment meters are favoured by the suppliers and customers. With gas, there is a strong inverse relationship between the provision of prepayment meters and disconnection rates (Ouseley 1988, Fig. 15). The decline in gas disconnections since March 1988 has been associated with an eightfold increase in prepayment meters: from 7,000 to 59,000 a year.

A family in debt and on IS can avoid disconnection by having money deducted at source from its benefit — known as 'fuel direct'. For these households, the average amount deducted from weekly benefit increased from £3.30 in 1977 to £8.80 in 1988 (Hansard 20 Dec. 1979, WA col 375-66 and 15 Dec.1989, WA col 856). This is a 40 per cent rise in real terms in eleven years, implying that fuel debts are getting bigger and fuel direct deductions represent a larger proportion of IS. When a claimant household has both fuel direct and HB deducted by the DSS at source, budgeting for other needs becomes increasingly difficult. The main energy suppliers and housing authorities obtain their payments, but the effect is to shift the manifestations of poverty into another area, such as inadequate nutrition, clothing or access to transport.

By combining the evidence of fuel debts (as in Table 3.2) with a proportion of those known to be offered repayment arrangements, Parker has estimated that in 1984 about 1.4m electricity consumers in England and Wales (7.5 per cent) and 1.0m gas consumers in Great Britain (6.5 per cent) had 'payment difficulties' (1985, pp.24-9). If Parker's figures are grossed up for the whole UK, this would mean that 1.6m electricity consumers had difficulty with their bills; the gas figure would not change as there is a minimal supply in Northern Ireland, to give a UK total for both fuels of 2.6m fuel debts in 1984. The overlap is not known, so it is impossible to estimate the total number of households with fuel debts, though 70 per cent of households have both fuels.

By 1990, 1.9m gas credit customers were repaying a debt through a payment arrangement and this represented a doubling of households over the period 1985-90 (British Gas statistics). The only possible double

Table 3.2 Households in difficulties with paying fuel bills, England and Wales for electricity, Great Britain for gas, 1970-89 (000s)

	Disconnected	Avoided disconnection as assisted through Code of Practice		Total†
		Prepayment meter installed	On fuel direct*	
1970	146.6			146.6
1971	145.3			145.3
1972	143.5			143.5
1973	146.7			146.7
1974	153.6			153.6
1975	176.2			176.2
1976	153.3			153.3
1977	123.2	99.5	90.0	312.6
1978	121.9	127.4	90.0	339.3
1979	124.0	114.7	63.0	301.7
1980	164.5	111.1	94.0	369.6
1981	140.7	170.8	143.0	454.5
1982	122.6	236.3	188.6	547.5
1983	120.1	262.2	231.8	614.1
1984	126.6	246.9	261.0	634.5
1985	137.3	240.1	300.0	677.4
1986	145.6	252.0	310.5	708.1
1987	150.1	234.3	301.0	685.4
1988	126.6	192.0††	304.0	622.6
1989	89.4	240.0††	212.0	541.4
Total**		2,527.3		

Notes:
† There may be some overlap between groups — perhaps 10 per cent (Conaty, pers comm), but the total is still given to identify the trend.
* A deduction is made from SB at source to cover both fuel arrears and current consumption.
** It is not appropriate to total the other columns as they represent temporary events.
†† Provisional, perhaps low, figures for electricity.

Sources:
Cols 1 and 2: Electricity Consumers' Council 1985, and press releases, National Gas Consumers' Council press releases, British Gas Corporation Annual Reports col 3: 1977-8 — Hansard 20 Dec. 1979, WA Col 375-6 (adjusted to exclude voluntary savers); 1979-80 — DHSS Annual Statistical Enquiry; 1981-4 — Social Security Statistics, Table 34.68; 1985 — Hansard 22 April 1985, WA col 344; 1986 — Hansard 9 Nov. 1987, WA col 98, Consumers' Councils' press releases and pers comm

counting with Table 3.2 could be the 150,000 claimants paying for gas on fuel direct. A payment arrangement is, therefore, the most likely outcome of getting into arrears with fuel payments and the annual total of households known to have fuel debt problems is rising rapidly. Depending on the growth in electricity payment arrangements, the overlap between households with payment arrangements and those in Table 3.2 and the numbers of families paying off debts for both fuels at once, the total number of households with fuel debts is in the range of 2.8-5.6m in 1990.

The trend in increasing fuel debts has continued despite decreasing real fuel prices (Figure 2.2). Furthermore, as 2.5m households with a history of fuel debts or on a low income have had prepayment meters installed, and are effectively practising self-disconnection if short of money, the number of households still experiencing fuel debts should be decreasing not rising.

Characteristics of fuel debtors

In his study of the Code of Practice, Berthoud found that three-quarters of both disconnected and debtor families contained children, whereas the elderly do not normally get into debt (Table 3.1). A high proportion of those with fuel debts were sick or disabled and a major reason for payment difficulties is a sudden change in circumstances, for instance as a result of illness or unemployment (Parker 1985, p.31). The reason for the drop in income may simultaneously be a cause of higher heating needs, because the home needs to be warmer for longer.

Finally, the majority of households with fuel bill problems live in rented accommodation (Table 3.1). As demonstrated below, about 70 per cent of the poor are tenants, so that fuel debts are not disproportionately associated with rented accommodation. However, tenants have less control over the environment of their home, and particularly its energy efficiency, than owner occupiers and are often required to purchase expensive warmth (Chapters 4 and 5). This could be the reason why a proportion of households in all income quintiles have disproportionately high fuel expenditure.

As explained at the beginning of the chapter, fuel bills can only provide partial information on the problem of fuel poverty — families with cold homes cannot be identified through their bills. The data on households with fuel bill problems have provided some evidence on the characteristics of those that suffer from fuel poverty. There are significant differences even between those with different types of fuel bill problems and no single social characteristic defines either group. At least 3m and up to 6m households have had difficulty paying their electricity and gas bills and are repaying a debt in 1990. Paying for fuel causes problems for a large and increasing number of households.

Defining poverty

Mack and Lansley found that people ranked 'heating in the living areas' first out of a list of 26 necessities, before an indoor toilet, damp-free home, a bed for everyone or three meals a day for children (1985, p.101). This means that heating in occupied rooms was the most important variable in a survey of public opinion in Britain in 1983, for defining a minimum standard of living. People give a high priority to adequate heating, if they have sufficient income and, therefore, fuel poverty is a clear indicator of poverty. Whilst the converse is not necessarily true, it is reasonable to assume that most of those in poverty are restricted in the amount of fuel that they can purchase and thus are suffering from fuel poverty. However, out of the plethora of definitions of poverty, which is the most appropriate definition to use? Defining and analysing poverty or social inequality is a complex process, which is not central to this study and which, therefore, is not undertaken in great detail. Furthermore, a pragmatic approach has to be taken because of the need to use published data to examine the growth of poverty, and by implication fuel poverty, and because of the possibility of policy responses which combine further help with an existing benefit. In addition, the housing conditions of the poor need to be identifiable to facilitate research into the role of capital stocks.

Relative deprivation

The approach recommended by Townsend is to consider whether the resources available to an individual are sufficient to enable participation 'in the activities, customs and diets commonly approved by society' (1979, p.31). This explicitly defines poverty in relation to changing standards and incorporates improvements in the living conditions of the poorest families, in line with rising general expectations. Townsend's research (1979) and similar work by Mack and Lansley (1985) provide useful snapshots of social expectations and conditions, at a point in time, in relation to a specified 'basket of goods'. However, the surveys are not easy to replicate in other years to reflect changing public perceptions and thus cannot assist in measuring poverty over time.

The concept of relative deprivation is appropriate for the assessment and definition of fuel poverty, because it allows for general standards, such as levels of warmth, to be included, which cannot easily be specified through amounts of income or expenditure. A simpler, but still relative, approach, would be to set the poverty level at a fixed (but arbitrary) proportion of average earnings. Thus, low pay has been defined by the Low Pay Unit as less than two-thirds of median male earnings (Smail 1985, p.2). Such definitions are difficult to relate to total family income, inclusive of benefits, on a daily basis. Thus, relative deprivation appears to be a sound concept, with considerable relevance to fuel poverty, but is difficult to use because of measurement and identification problems.

Income support and housing benefit

An alternative basis for a poverty datum line is in relation to the income level below which a household qualifies for a means-tested social security payment. (Universal benefits, such as the state pension and child benefit, are paid irrespective of income, and the numbers of recipients indicate demographic trends rather than the extent of poverty.) The fundamental principle underlying income support for the poor is that claimants can spend their money as they wish and it is invidious for government to specify amounts of money for particular purposes. Thus, IS scale rates are not based on an analysis of the cost of a basket of necessities and no attempt has been made by government to specify a minimum standard of living since Beveridge set the rates for National Assistance in 1948 (Andrews and Jacobs 1990, p.178).

If there is a real increase in the value of means-tested benefits, more people are eligible to claim and are apparently in poverty, although hardship has diminished. The converse can also occur. For instance, the amount by which IS is increased each year used to be linked to average earnings, but since 1980 it has been tied to the RPI, which has been rising more slowly. In addition:

> The way the RPI is constructed means that it will be closer to the inflation experience of the rich than of the poor... Between 1974 and 1982, the price index of the poorest 10% of households rose by 17 percentage points more than that of the richest 10%. (Fry and Pashardes 1986, p.64)

This is because 'a number of the items whose consumption is most heavily concentrated among the poor, such as coal and electricity, are also those whose prices have risen most dramatically' (ibid, p.37). Therefore, it may appear that benefits are retaining their real value, through the RPI, whereas the living standards of recipients are declining together with the number of claimants. However, the numbers of recipients of specific benefits do indicate the extent to which the government has accepted responsibility for providing support.

There are three main means-tested benefits (IS, HB and family credit). With IS, the assessment unit is an individual (plus dependants). Non-dependent members of the same household (for instance an unemployed 18-year old) are treated as separate income units for IS purposes. Many IS claimants are, therefore, members of someone else's household or in board and lodgings and not directly responsible for the payment of fuel bills or the condition of the house. For similar reasons, these non-householders do not qualify for HB (Cohen and Lakhani 1986, pp.32-4, 163). Thus, there is only a partial overlap between receipt of IS and HB, as shown in Figure 3.3. The receipt of IS by non-householders means that this benefit provides a confusing basis for fuel poverty analysis and is avoided wherever possible.

HB covers some or all of the housing costs for IS and Family Credit claimants and other low-income households and is, therefore, the most

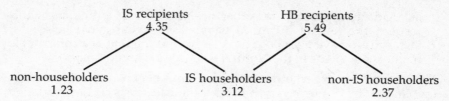

Figure 3.3 Relationship between receipt of housing benefit and income support, GB 1988 (million households)

Sources: SSS 1989, pp.330,350

widely spread means-tested benefit. For households on IS 100 per cent of housing costs are covered normally, but for other households the support is tapered in relation to income. In all cases, support is related to the actual costs incurred, rather than to some hypothetical sum or estimate. As Figure 3.3 shows, less than 60 per cent of HB recipients were IS claimants, which demonstrates the quite different populations covered by these two benefits.

Before 1988, the number of HB claimants was rising, reflecting the increases in SB claimants and the real cost of rents — between 1978-9 and 1983-4 rents rose at double the rate of inflation (Mack and Lansley 1985, p.6). The 6.8m recipients of HB in 1987 represented approximately 34 per cent of all private households in GB. However, the April 1988 revisions to the social security system originally resulted in 1m households losing all entitlement to HB (Andrews and Jacobs 1990, p.36), as reflected in the total of 5.5m households shown in Figure 3.3. Subsequent revisions may have increased the number of recipients again.

One of the main disadvantages of means-tested benefits is that they are not taken up by all those eligible to receive them. With HB, the proportion claiming was estimated to be 77 per cent in 1984 (SSS 1988, p.276). These non-claimants represent a further 1.9m households, to bring the total to 8.3m who could have been claiming in 1984 — a remarkable 42 per cent of all households in GB (no UK figure is available). All of these households are on an income, which is deemed insufficient by government to cover revenue expenditure on their house. (Capital expenditure on repairs and improvements is not covered by HB.) This is partly because:

> In the DHSS ... there was no branch — indeed, no official of any kind — whose job it was to worry about 'take-up' or to report regularly on the Department's success in fulfilling its legal obligations to get benefits to the people who were supposed to have them. (Donnison 1982, p.42)

Low-income households receive support with housing costs through HB, because of the 'mismatch between their resources and the cost of housing' (Cmnd. 9520 1985, p.v). The reason for excluding housing costs was that:

> housing costs varied so greatly from household to household that to include housing in the scale rate would leave many householders with

an inadequate amount for other needs. Housing is unique in this respect: the rent of a council flat in Kensington and Chelsea for instance is on average about twice as high as the rent of a comparable flat in some towns in the North. (Treasury 1985, pers comm)

The government expect other, everyday items of expenditure, including fuel covered by IS, not to vary as much as housing costs and any variation is thought to be in relation to family composition. Whilst the latter is an influence, the much greater variations that are caused by the energy efficiency of the home are demonstrated throughout the following chapters, and have been recognized by Parker:

> The SB system currently fails to acknowledge that fuel costs vary substantially from household to household; consequently payment for fuel is a considerable problem for many claimants. An income support system which was responsive to these variations ... would be of great value. (1985, p.34)

There is no assistance with fuel costs now that HA have been abolished. Thus, the government's underlying perception appears to be that because housing costs vary so much assistance has to be individually assessed; fuel costs are a matter of personal choice and part of the general budget. The individual matching of needs and means that is demonstrated by receipt of HB is, therefore, accepted as an appropriate indicator of poverty for the purposes of fuel poverty analysis.

Dependence on the state

Receipt of a benefit does not indicate the extent to which a family is dependent upon the state. By examining the sources of income, it is possible to assess the role of government as a provider for those on the lowest incomes and, thus, the extent to which the state is determining their standard of living. The definition of dependence used is that for identifying poor pensioner households:

> A retired household mainly dependent upon state pensions ... is one in which at least three-quarters of the total income of the household is derived from national insurance retirement and similar pensions, including benefits paid in supplement to or instead of such pensions. (FES 1985, p.92)

This definition of 75 per cent of income, or more, coming from the state has been extended in this study to define dependence upon the state for all households (Table 3.3).

The Table shows the proportion of gross income received in benefits by households in the two lowest income quintiles, whether the benefits were universal or means-tested. Whilst some of these households have no income, other than benefits, many have nominal amounts of income. In 1982, the majority of the original income of the lowest quintile, before receipt of cash benefits, came from occupational pensions, investments and annuities, and less than 1 per cent was earned income (*Social Trends*

Table 3.3 Proportion of gross* income received from benefits and other sources
of income for low income households, UK 1974-86 (%)

	Lowest income quintile		Next income quintile	
	Earned and unearned income	Cash state benefits	Earned and unearned income	Cash state benefits
1974	19	81	81	19
1975	18	82	78	22
1976	17	83	73	27
1977	14	86	82	18
1978	14	86	72	28
1979	8	92	67	33
1980	8	92	67	33
1981	8	92	63	37
1982	5	95	57	43
1983	4	96	53	47
1984	3	97	51	49
1985	4	96	51	49
1986	3	97	54	46

Notes: Pre-1974 information is not available on a comparable basis.
* Before deduction of taxes

Sources: 1974-82 *Social Trends*, Table headed 'Redistribution of income through
taxes and benefit'; 1983-6 *Economic Trends*, December 1984, p.94; July 1986,
p.102; November 1986, p.101; December 1988, p.91

1985, p.88). Only 2 per cent of households in this quintile contained
someone who had been employed for the whole year. Whilst it is evident
from Table 3.3 that an increasing number of households are dependent
upon the state, it has not been possible to calculate the exact proportions:
the figure is estimated to be just over 20 per cent in 1973, rising to about 30
per cent in 1986. As the number of households in the UK was also
increasing, this represented a growth from 3.9m families dependent
upon the state in 1973 to 6.4m in 1986 and 6.7m in 1990.

As the data are for the two lowest income quintiles, it appears that
dependence upon the state and a low income are broadly synonymous.
Therefore, information about the 30 per cent of households dependent
upon the state can be deduced from data on the 30 per cent of households
with the lowest incomes. This Table certainly confirms that an increasing
proportion of families are dependent upon government for their income
and their standard of living.

A working definition of poverty and expenditure on fuel

In 1987 and 1988, a third or a quarter of all households received HB
(depending on the year taken) and about 30 per cent of households

depended upon the state, so these two methods of defining poverty affect similar proportions of families. It appears that the 30 per cent of households with the lowest incomes are a proxy group for the other two definitions, with, hopefully minor, dissimilarities. Therefore, reference to 'the poor' or to 'low-income households' in this book refers to the 30 per cent of households with the lowest incomes, whether in the UK or GB. This is statistically convenient when extracting information from the FES. The households in receipt of HB are readily identifiable for targeting policies. Thus, between them, these three definitions provide a basis for data analysis, a benefit to which assistance can be linked and the justification for government involvement. It has to be emphasized that although the overlap between these three groups is undoubtedly considerable, it cannot be identified with any certainty.

Now that the poor have been defined it is possible to identify their fuel expenditure in comparison with that of other families (Table 3.4). Although the poor spend less on energy in absolute terms than better-off families, this money represents 10.0 per cent of their expenditure in comparison with only 4.4 per cent for higher income households.

Table 3.4 Household weekly expenditure on fuel by income, UK 1988 (£)

	30% of households with lowest incomes	70% other	Average
Electricity	4.21	5.24	4.93
Gas	2.85	4.71	4.15
Coal and coke	0.92	0.78	0.82
Other	0.50	0.61	0.58
Total fuel expenditure	£8.48	£11.34	£10.48
All household expenditure	£84.56	£255.77	£204.41
% on fuel	10.0	4.4	5.1

Source: Based on FES 1988, Table 5

Isherwood and Hancock (1979) found that 18 per cent of households spent twice the median on fuel in 1977; the median figure for 1988 is not known, but 30 per cent of households were spending nearly twice the average of 5.1 per cent, implying that disproportionately high fuel expenditure is also a worsening problem.

Because the 30 per cent of households with the lowest incomes are known to be dependent upon the state for at least 75 per cent of their income, it is possible to calculate the amount of money that they spend on fuel that has come from benefits. In 1988, for two-thirds of these households, 97 per cent of their income was from the government (based on Table 3.3), and the level of support was about 75 per cent for the remainder. As there were 21.9m households in the UK in 1988 and using weekly fuel expenditure data from FES 1988 (p.21):

4.38m households @ £8.15 per week × 52 weeks × 97% = £1,800m
2.19m households @ £9.41 per week × 52 weeks × 75% = £ 800m

This calculation reveals a total expenditure on fuel by poor households in excess of £2,600m out of state-derived income in 1988. This is 89 per cent of the £2.9bn spent on fuel by these households out of all income.

Government capital expenditure on energy efficiency improvements to the homes of the poor is estimated (Chapter 4) to be infinitesimal in comparison with the substantial sums of money being spent through imputed revenue expenditure on fuel. The government is providing a large amount of money for the fuel bills of the poor but is not, concurrently, investing to ensure that the poor have energy efficient homes. The ability of the poor themselves to redress this marked imbalance between capital and revenue expenditure is assessed next.

Characteristics of the poor

Savings, debts and access to credit

Income levels provide part of the information about a household's financial situation, though the additional resources represented by savings or access to credit are also important, particularly if capital expenditure is to be considered. Debt is used in its popular sense to mean default in payment and indicates problems with household expenses for whatever reason. The number of families in debt or with multiple debts is growing, but as 'there are no comprehensive national statistics on the scale of consumer debt problems' (NACAB 1986, p.5), it is not known how many of those with fuel debts have other debts as well.

All recipients of a means-tested benefit (for instance HB) have, by definition, limited capital. The majority of SB claimants were known to have no capital assets (i.e. savings) with the proportion rising from 61 per cent in 1978 to 69 per cent in 1987 (SSS Table 34.55). Pensioners are virtually the only group with savings, whereas 53 per cent of couples with children are in debt (Berthoud 1984, p.A6a). In 1982, more SB claimants owed money for fuel debts than for any other reason (Berthoud 1989, p.29). The recent growth in mortgage arrears and of debt generally create problems for a broad spectrum of incomes. However, the poorer people are, the more debts they face (Berthoud and Kempson 1990, p.14), though, as Parker reported:

It appears that those who have the lowest incomes are actually less likely to use credit than those with higher incomes... (which) gives the lie to popular myths of the 'feckless poor' spending all their state benfits on the hire purchase of cocktail cabinets and colour televisions. (1985, p.6)

The NCC report a similar finding in 1987 (1990, p.135). This is not as contradictory as it sounds. Low-income consumers are more likely to be in debt with their regular payments, for instance for fuel, rather than because they have taken on credit purchases specifically. For instance, they have, unfortunately, to purchase gas and electricity on credit, because the industries refuse to provide prepayment meters on demand. One reason for the small numbers of low-income families using credit is that 'pensioners, as a group, have low average incomes *and* are less likely to have used credit in any form'. The heaviest users of credit tend to be the young, regardless of income, probably because of 'the expense of furnishing the home as well as supporting young children' (Parker 1985, p.6). Many IS/SB families, particularly those with children, are living beyond their means and may have a weekly expenditure greater than gross income.

Low-income families, headed by skilled workers, that do use credit are likely to use finance company loans with interest rates varying from 30-40 per cent; other families, in semi-skilled manual occupations, are more likely to use check traders and money-lenders (very common forms of low-income credit) with interest rates at 50-600 per cent (Parker 1985, p.5; Conaty 1987, pers comm; Ashley 1983, p.78). More than half of socio-economic groups D and E in 1985 had no current bank account and they were predominantly women (Golding 1986, p.59), confirming that many of the poor have limited access to cheap credit.

Without savings, perhaps in debt, in difficulty with current bills, and having access only to expensive credit, the poor are restricted in their opportunities for capital investment in their properties. If they do undertake capital expenditure, the rate of return on the investment would have to be higher for low-income families than for other households, in order to provide an appropriate investment and use of money from the family's perspective.

Not all income from employment is reported, so that some low-income families could be poor on paper, but not in practice. In a report by the OECD, the size and locus of the black economy (concealed employment) was considered:

> there is little evidence... that would sustain a belief that concealed employment is especially important among those recorded as unemployed in the regular economy. (quoted in the *Guardian* 24 Sept. 1986)

Similarly, in a survey of 676 people, Pahl found 27 people (4 per cent of respondents) 'who acknowledged receiving extra money informally', but of these only one was an unemployed male and eight were housewives (1985, p.13). Pahl concluded that 'for the unemployed who most need money, it is too dangerous to take it', because of the risk of losing their benefit (ibid, p.14). Therefore, the recorded income for most people dependent upon benefits is likely to be a fair indication of their actual income.

Social characteristics

The majority of the poor are at home, not out at work, because they are pensioners, single parents, unemployed, sick and disabled, though there are significant numbers of low paid (for instance those in receipt of Family Credit). The numbers of poor in the future depend upon demographic, social and economic trends.

Life expectancy has been increasing steadily, so there are both more old people and they live longer. The result is a strong association between age and the depth of poverty: 54 per cent of households headed by a person aged 65-75 years and 72 per cent of households where the head is aged 75 years or more are poor (FES 1988, p.6). Pensioners, especially the very elderly (over 75 years), are likely to remain the largest group of the poor (Table 3.5). This is a particularly British problem as, within the European Community, the UK 'had the largest proportion of the population aged 60 or over' in 1985 (*Social Trends* 1988, p.34).

According to Haskey (1986, p.9), between 1980 and 1984:

the number of families with dependent children decreased by 2 per cent

the number of lone parent families   increased by 13 per cent

the number of lone parent families on SB   increased by 55 per cent

This confirms that lone parents are at great risk of poverty. Whether this risk will rise further is not certain as the rate of increase in the number of divorces has slowed down, whereas the number of single parents who have not married is increasing (Haskey 1986, p.10).

The number of people who are registered as unemployed peaked in mid-1986 and was down to 1.6m by early 1990, though part of the fall is due to changes in eligibility. At the end of 1990 unemployment levels were rising again. Households headed by an unemployed person constitute the second largest group of claimants in 1987 (HC 378-II 1990, p.xi). The number of long-term unemployed is still high: by January 1989, 0.8m claimants had been unemployed for more than a year (Hansard 10 April 1989, WA col 369).

The number of poor families headed by someone in work is rising. One reason is that between 1978-9 and 1983-4 the real increase in earnings for workers on two-thirds of average incomes has been less than that of workers on 1, 5 or 10 times average earnings: 5 per cent, 6 per cent, 22 per cent and 54 per cent respectively for a married man with two children (Hansard 23 Nov. 1983, WA col 167-8). As shown above, many low-paid workers and their families are poor and have fuel bill problems, so the definition of poverty should allow low-paid workers to be included.

There is, therefore, little prospect of a decline in the numbers of people in poverty. With increased longevity and a growth in the long-term

Table 3.5 Household characteristics, by head of household, UK 1988 (% of group)

	30% of households with lowest incomes	All
ECONOMIC STATUS:		
Pensioners	58	26
Non-pensioners:		
Unemployed/unoccupied	28	13
Workers	14	61
Single parent	10	4
All families with children*	18	35
TENURE:		
Local authority tenants	57	25
All renting	69	35
Own outright	26	24
Purchasing	5	41
Numbers	6.6m	21.9m

Note:
* Overlaps with other groups

Source: FES 1988, p.6

unemployed, poverty is increasingly a long-term state. Much of the growth in the numbers of poor families has come from the unemployed, lone parents and the low paid, and each of these groups has more dependants in the family than there would be in a pensioner household. Therefore, more children, more people and younger people are being affected by the growth in poverty.

Tenure and housing conditions

The majority of poor households are local authority tenants (Table 3.5), representing in turn the majority of all public sector households. The concentration of low-income households in council housing — its 'pauperization' — has been a distinctive trend, at least since 1953 (Forrest and Murie 1987, pp.20-2; Merrett 1979, p.201). Only since 1982 has the number of households paying off a mortgage risen above the numbers renting from a local authority or new town (GHS 1987, p17). The particular importance of the whole rented sector is clear: 70 per cent of all poor households are in rented accommodation. The other large group of the poor are outright owners, despite having a capital asset. This predominance of poor households in rented property is reflected, wherever possible, in the analysis in later chapters.

Education and behaviour

So far in this chapter, the emphasis has been on levels and sources of income and on the characteristics of the poor that are linked to these factors. However, there are other aspects of life that affect the way people, particularly the poor, use heating or fuel.

Knowledge and understanding

According to Williams, 'Some homes are badly heated and unhealthy because the occupants make them so' (1986, p.5), due to a lack of knowledge and a failure to 'manage' the building in the way that designers had anticipated. If people do not understand the technology of energy use in the home they describe what happens in terms that fit with their existing beliefs and knowledge, 'even though it does not match the physical situation as understood by the scientist' (ibid, p.4). This can inhibit the more efficient use of energy. For instance, if cold is thought to come into the house, like draughts, rather than heat going out, there is a misunderstanding of the role of insulation: 'I don't need insulation, the house is warm enough' (Bagshaw 1981, p.76).

In some cases problems are caused not because of a failure to understand the technology, but because the technology is based on assumptions that are inappropriate to the user's values. The belief that 'fresh air is good for you' can mean that a house is well ventilated in the morning, which is a wasteful strategy with electric storage heating and means that expensive supplementary heating is more likely to be needed in the evening (Crawshaw et al. 1985, p.287). This form of 'misuse', according to McGeevor, is not 'a symptom of incorrigible ignorance', but indicative of 'the degree of social distance between the "expert" and the ordinary user' (1982, p.105). Whilst people can learn how to operate new heating systems, most of their information normally comes from other users: technical information is 'filtered by the social process' to make it available to the lay person in the form of 'folk wisdom' (ibid, p.106).

With traditional heating methods, such as coal fires, there is a considerable store of folk wisdom, so that successful performance is achieved. With new heating technology, this folk wisdom has not accumulated and, without better instruction and feedback, is unlikely to develop accurately (ibid, p.106):

> It's too hot and I don't know how to make it right. I'm not used to central heating. I turn it on and off but I don't really understand it. I wish someone would come and explain it. (A partially blind council tenant of 87, living alone, quoted in Durward 1981, p.39)

Another reason for the inefficient use of energy is ignorance of relative fuel prices and running costs. Knowledge of the latter was related to age in working-class women, and was greatest for women under 36 and least for women over 60 (Bagshaw 1981, p.59):

> My wife thinks she's been economising and it should be less but it

always comes in about the same. Maybe we economise on the wrong things. (An owner occupier, quoted by Durward 1981, p.38)

A survey for the Electricity Consumers' Council, established that only 15 per cent of consumers know even vaguely how much a unit of electricity costs (1985). Without knowledge of actual and relative costs, strategies to reduce running costs are likely to be inappropriate:

I sit with my curtains back to save fuel — saves putting the light on... I do cut my electric if I can because I have big gas bills. (An elderly lady with gas CH, quoted in Durward 1981, p.39)

Relative fuel costs are important in determining the cost of warmth for different households. The emphasis placed on fuel pricing by government since 1976 assumes that consumers are aware of the original cost of the fuel. After the price of gas rose substantially between 1979-82, only 65 per cent of people perceived gas as cheaper than general tariff electricity (Heslop and Sussex 1984, p.5), even though, in delivered energy units, it was a quarter of the price (Chapter 5). If there are widespread misconceptions about relative and actual costs, as appears likely, individual and national strategies to use energy efficiently will be unsuccessful — the last three quotations used above come from people trying to economize. Many consumers, therefore, probably have unnecessarily large fuel bills because of their lack of knowledge and poor understanding of energy use.

Literacy and information

Misconceptions about energy use in the home are not restricted to low-income households, but obtaining better information may depend upon the ability to read. However,'it is estimated that about 6% of the adult population are functionally illiterate... (with) such limited reading and writing skills as to seriously affect almost every aspect of their lives' (Adult Literacy and Basic Skills Unit). There is no suggestion that all poor people have reading or writing difficulties, but those that are functionally illiterate are more likely to be poor. Problems with filling in forms is cited by Deacon and Bradshaw as one of the reasons for failure to claim benefit (1983, p.136).

With limited literacy, improving their knowledge of energy use in the home is difficult. Crawshaw and Dale (1981) showed that the operating instructions provided for the Electricaire heating system given to council tenants required a reading age of 15+ years. A reading age of at least 10 years was required for most leaflets on energy conservation, because of the use of long sentences or words, or technical expressions such as 'lag' or 'thermal insulation'. A further obstacle to comprehension is that many elderly working-class women understand the word 'energy' to mean personal energy, rather than a fuel (Bagshaw 1981, p.69). Therefore 'many of the available (consumer education) materials are not readable at a suitable level for those needing ... basic information' (ibid, p.136).

Verbal advice on appliance usage comes mainly from the electricity and gas showrooms (apart from folk wisdom), though Meyel found that low-income households in London only used the showrooms for fuel bill queries or to purchase appliances. In most cases, the only direct advice on energy efficiency had come from double glazing salesmen (Meyel 1987, p.40). The importance of verbal communication to overcome language and literacy barriers is stressed by research funded by the EEO (EEDS 56 1988, p.3).

Recognizing that a good reading ability helps people to achieve the efficient use of appliances, read meters and bills and understand the relative costs of different fuels, does not demonstrate how much increase in energy efficiency could come from improved levels of knowledge. According to Williams, what people 'do in their homes can denigrate the most perfect environment, or make the most hostile conditions acceptable' (1986, p.1). This is probably rather extreme, as American research has shown that only 18 per cent of the variation in usage between households can be accounted for in the USA by 'lifestyle' factors, such as the amount that windows are opened, whether the time clock is used, as well as temperature setting (Sonderegger 1978, pp.201-28). This research finding has been criticized because the respondents were from a fairly homogenous social group, although living in a variety of houses. The effect of lifestyle and the other factors influencing energy use variations are shown in Figure 3.4.

To ensure that people can control their home environment adequately advice needs to be specific to the individual. Any family (whether low-income or not) with a poor understanding of energy use may not realize it lacks knowledge, and, even if it does, if it has a low level of literacy, may not be able to do anything about it. Therefore, even though people's behaviour may be partly responsible for the inefficient use of energy in the home, they require outside assistance in order to remedy the situation.

Money management

There is one final area where people's understanding and behaviour could be exacerbating the extent of their fuel poverty and that is through the way they manage their money. Fuel debts could be caused by a lack of money when bills are due, rather than a shortage of money overall: the image of the feckless poor. Evidence to prove or disprove the concept is difficult to find, though:

— for the poor, fuel is the third most important item, whereas expenditure on less essential items, such as tobacco and alcohol, is much lower;

— 'It is sometimes said that the problem with the SB scheme is not that the rates are inadequate but that those who are dependent on the scheme cannot manage. However, a number of the interviewers, all experienced social workers and familiar with the families, still found themselves shaken by the grind and stress that the poverty of the scale rates imposes. Managing constantly and effectively over a long period (a year or more) requires a personal security and strength' (Burghes 1980, p.73).

— over 40 per cent of consumers with fuel debts (whether disconnected or not) thought slot meters the most convenient method of paying for gas and electricity, whereas only about 10 per cent of all households actually had them in 1980 (Berthoud 1981, pp.222, 227);

— a study of working-class women found that those with little knowledge about the cost of electricity were most likely to have a slot meter or weekly savings plan; they were cautious and planning their payments (Bagshaw 1981, p.63-5).

These last two examples demonstrate the ways in which the supply industries could assist low-income consumers through the provision of

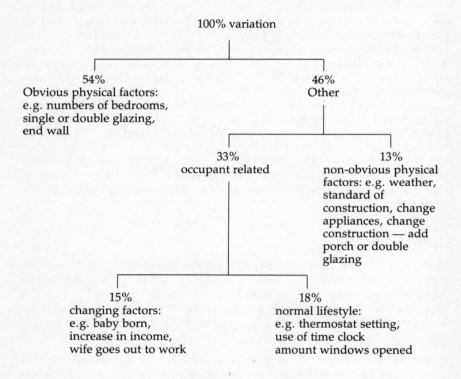

Figure 3.4 Causes of variation in energy use between non-indentical houses*, USA

Note:
* A variety of house styles were included

Source: Sonderegger 1978, pp.201-28

appropriate payment methods. The extent to which fuel poverty is exacerbated by the problem of three-monthly credit for gas and electricity is not known. One of the factors that highlighted the problem of fuel poverty in the early 1970s was the increase in disconnections, and these could have been caused, at least in part, by a rapid phasing out of prepayment meters; these meters are now generally available only to consumers with a proven history of fuel debts. Neither the gas nor electricity industry design schemes specifically to assist the poor. As one electricity board official stated to Berthoud, 'none of the easy payment schemes are directly aimed at low-income consumers' (1981, p.37), nor are they intended for payment of accounts overdue (Electricity Council leaflet 1985). New coinless prepayment meters and proposed multi-tariff meters are designed to benefit the industry more than the poor consumer.

Conclusions

Income-related evidence of fuel poverty has continued to grow at a faster rate than the growth in the numbers of poor. For instance, the number of households who were disconnected, on fuel direct or had a prepayment meter installed to avoid disconnection rose by more than 60 per cent between 1977 (the first year of full statistics) and 1989, whereas there was only a 40 per cent increase in the number of households dependent upon the state for the whole period 1970-89. The detailed evidence on fuel bill problems substantiates the findings of the previous chapter — fuel poverty is a continuing and growing problem. As this increase has occurred despite both increases and decreases in real fuel prices, it is possible to deduce that the causes of fuel poverty are more stable than fuel prices and are impinging with increasing severity on the poor.

The poor are pensioners, sick and disabled, lone parents, unemployed and low-paid workers, but within each of the first three categories there are substantial proportions that are not poor. Therefore, there is no characteristic or social indicator that would enable additional help, for instance fuel allowances, to be targeted effectively, except through an existing benefit. This analysis starts from the assumption that there is some (imperfect) link between poverty and fuel poverty, because the latter is based on the definition 'an inability to afford adequate warmth'. The most appropriate definitions of general poverty identified are dependence upon the state, with at least 75 per cent of all income being received in the form of cash benefits, or receipt of the means-tested housing benefit. Both of these categories include about 30 per cent of all households. Therefore, the poor are defined as the 30 per cent of households with the lowest incomes. Not only are these households dependent upon the state for their standard of living, but the majority have no savings, or very limited amounts. Furthermore, over 70 per cent of them live in rented accommodation. This population, therefore, is unlikely to undertake capital investment in the energy efficiency of their home because of legal or financial constraints, or both: they are beyond the reach of market forces. Thus, the poor are severely constrained in their ability to invest in capital improvements to the properties they occupy. The government's responsibilities to this group of people, therefore, covers both income support and capital investment.

The large proportion of expenditure that goes on fuel indicates why fuel price rises are particularly regressive for the poor. The 6.6m poor households spent, on average, just over £8 a week on fuel in 1988. Of this expenditure, a total of £2.6bn came from state-derived income and yet this is still insufficient to prevent rising fuel poverty. As has been stated, at least the bottom 20 per cent of households have experienced a real cut in income between 1979 and 1987, so the growth in dependence upon the state is not as a result of a more generous benefit system. It is evidence of considerable additional poverty affecting both more households and each household more deeply. As Professor Peter Townsend has stated:

> It is right that we should be gloomy about conditions in our own country. They have deteriorated. Whether we are talking absolutely or relatively, there are *more* people experiencing worse conditions today than when I was young and tramping the streets of the East End of London in the 1950s and 1960s. (1990, p.9)

4 Retaining heat

Now that the poor have been identified as a group, it is possible to look at the first part of the energy efficiency of the homes that they occupy — the ability of the building fabric to retain heat. The second part — the energy efficiency of the heating system, and the creation of warmth — is considered in Chapter 5. This chapter begins with an investigation into how variations in building energy efficiency occur and what range exists within the present housing stock. In order to clarify the causes of fuel poverty, it needs to be established whether any variations in thermal insulation levels result in detriment to the poor, so that they have to pay more per unit of warmth than other households.

The emphasis here is mainly on technical aspects of energy efficiency, with an introduction to economic issues. The latter are expanded further in Chapter 9 with the consideration of cost effectiveness in the context of government policy. Some of the social and behavioural dimensions to the demand for warmth, such as levels of occupancy and comfort temperatures, are examined in Chapter 6 to define a target heating standard for low-income households.

The energy efficiency of the building fabric in the present housing stock reflects two effects of government policy, both of which were mentioned in Chapter 2. The first is the failure to recognize the importance of warmth as an element of satisfactory housing and to frame building and environmental health standards accordingly. The second is the desire to reduce energy consumption as a result of the first oil crisis. The effects of past government policies are explored further with an analysis of the impact of Building Regulations and government involvement in the repair, improvement and insulation of houses. The role of government is particularly important because improvements to the energy efficiency of an existing building can be achieved only through capital investment and it has already been established that the poor are largely dependent upon the state for both their income and capital expenditure on energy efficiency improvements. Of the four factors (COWI 1-4) that represent the potential for energy efficiency investment, the price of the fuel used (COWI 4) has effectively been excluded from fuel poverty policy, for macro-economic reasons. Therefore, the opportunities for improvment are limited to insulation, ventilation control and type of heating system: two of these three factors are covered by this chapter.

Measurement matters

The rate of heat loss from an unoccupied house is the sum of the conduction of heat through the building fabric and the convection of heat through the gaps in the structure. The occupants introduce other factors, such as the amount windows and doors are opened, as mentioned in the previous chapter. Measuring or predicting the rate of heat loss in an empty building is an imperfect art as it is difficult to adjust the figures for such realities as disrepair in the building or poor standards of construction.

Rate of fabric heat loss — COWI 1

To calculate the rate of fabric heat loss, data are required on the dimensions of the building and the materials used in its construction. Heat is conducted through the materials used in buildings at different rates. A transmittance value, or U value (see Glossary), is calculated for each type of construction to reflect the combined rate of heat loss for all the component parts. Thus, a solid brick wall plastered on the inside has a U value of 2.1 W/m^2 °C, which means that a square metre of wall will lose 2.1 W for every degree of temperature difference between inside and outside. A double skin of bricks, with a cavity between and plaster on the inside, has a lower U value of 1.5 W/m^2 °C, and provides 40 per cent more resistance to heat loss (Markus and Morris 1980, p.304). Thus, lower U values represent better levels of insulation.

The materials and methods used in constructing houses have varied over time, though the diffusion of those new building techniques that improved the thermal insulation, such as putting felt under the tiles and building cavity walls, is not well understood. For instance, Muthesius considers that 'cavity walls were ... fairly frequent by 1900' (1982, p.65); Leach *et al.* were advised that cavity walls were introduced in the 1930s on a wide scale (1979, pp.97-8), though the practice was described as 'new' in a 1938 publication (Woollcombe, p.12). Furthermore, many of the industrialized buildings (see Glossary), such as the 1950s-60s blocks of flats, incorporated solid walls. Before 1965, the standards required in the construction of dwellings were determined by local by-laws which mainly addressed the stability of the structure and health considerations. Thus, the construction details and rates of heat loss of pre-1965 homes are best determined by individual surveys. Some indicative U values for the major elements of a building, by date of construction, are given in Table 4.1.

The Building Regulations stipulate the thermal insulation standards (amongst many other directives) to be achieved in certain elements of the building — roof, external walls, ground floor — and are mandatory in new buildings and extensive conversions. Other alterations can be undertaken without any improvement to thermal insulation:

> rehabilitation schemes need not comply with current building regulations on thermal insulation. As a result condensation and very

high heating costs are a constant complaint of many tenants. (Levitt and Burrough 1979, p.22)

The 1965 Building Regulations, the first to cover the whole country, largely confirmed existing standards, but also required the equivalent of 20mm of insulation in the loft (Table 4.1). From 1974 onwards, the amount of thermal resistance required by the Regulations has increased: because the U value is affected by the thickness of the materials being used, a change from 1.2 to 0.6 does mean a doubling in the efficiency with which heat is retained by that element. The 1974 amendment represented a significant increase in thermal insulation standards for new homes, many of which were fitted with cavity wall insulation to achieve a U value of 1.0 for external walls. The additional wall insulation required in the 1982 standard has increased the prevalence of cavity wall insulation further. The most recent Building Regulations, effective from 1 April 1990 in England and Wales, are still below the standards required in other

Table 4.1 Rate of fabric heat loss by date of construction

	UK Pre-1944†	1945-1965††	1966-1974	1975-1981	1982-1990	1990-	Sweden 1984
U VALUES (W/m² °C)							
External wall	2.10	1.50	1.70*	1.00*	0.60*	0.45*	0.17
Windows	5.6	5.6	5.6	5.6	5.6	5.6	2.0
Suspended floors	1.42	1.42	1.42*	1.00*	0.60*	0.45*	0.20
Roof	2.0	1.9	1.42*	0.60*	0.35*	0.25*	0.12
Roof insulation level implied (mm)	0	0	20	50	100	150	250
Fabric heat loss** (W/m² °C)	370	340	325	240	195	175	65

Notes: From 1966, the age bands reflect the dates when Building Regulations took effect, rather than when they were published. Before 1965, the choice of dates is somewhat arbitrary; data on the housing stock by age are usually, but not always, subdivided by the two World Wars, a convention followed here.
† If the dwelling has solid walls and no felt under the tiles
†† If the dwelling has cavity walls and felt under the tiles
* Required in the Building Regulations, broadly similar throughout UK
** Based on a standard mid-terrace house of 96m²

Sources: Pre-1944 and 1945-65: Based on Markus and Morris 1980, pp.304-9; and NEDO 1974, p.29; 1966-74: Building Regulations 1965, Statutory Instrument No. 1373 (1965); 1975-81: Building Regulations 1972, Second Amendment 1974; 1982: Building Regulations 1976, Second Amendment 1981; 1990: Davidson 1990, p.17; Sweden 1984: IEA 1989, p.57 Roof insulation levels: based on Eurisol 1984

countries, for instance Sweden. The Secretary of State for the
Environment is reportedly considering higher Building Regulation
standards already, partly to reduce greenhouse gas emissions.

The effects of Swedish and other international building standards on
the energy efficiency of houses are shown in Figure 7.6. The Building
Regulations have never covered all parts of the external fabric — for
instance, there is no specific limitation on the heat loss rate through
windows — and therefore they are concerned with a diminishing
proportion of the total heat loss (Figure 4.1). The new regulations do
allow 'trade-offs' so that a lower level of insulation in walls, roof or floor
can be accepted in return for the installation of double glazing.

To obtain the rate of fabric heat loss for a building requires each of the
main components (roof, floor, wall, windows) to be measured and the
area of each component to be multiplied by the appropriate U value.
Amongst other important parameters, therefore, the estimated fabric
heat loss takes into account the number of external surfaces, the extent of
glazed exterior surface and building age. Table 4.1 includes, for a typical
mid-terrace house, the effect of the different standards on the rate of
fabric heat loss: houses built prior to 1974 lose heat at least 86 per cent
faster through the fabric than those built to the present Building
Regulation standards. Substantial improvements in the thermal
insulation standards required in new buildings by successive revisions to
the Building Regulations have had the effect of widening the differentials
with older housing.

Ventilation losses — COWI 2

Heat is lost when warm air escapes from the building through cracks,
open windows and doors. The natural ventilation necessary to prevent
condensation, remove odours and to supply combustion is allowed for
through purpose-provided opening windows, air vents and chimneys.
Other unintended openings occur as cracks in the building fabric, either
left during construction, such as around window and door frames, or
caused by age, subsidence or shrinkage. These adventitious openings
vary between dwellings and over time in an unpredictable way, but can
be, in UK dwellings, much larger than those of purpose-provided
openings (Etheridge and Nevrala 1978, p.296).

The actual amount of air movement through the building depends
upon the way the occupants use windows and doors and on the pressure
difference between the inside and outside of the dwelling as a result of
prevailing winds or temperature range. Unfortunately, there is no simple
or inexpensive test or model available to quantify ventilation rates in
occupied dwellings, and thus very little evidence of the rates found in
British homes. Therefore, the estimated values for the rate of air
movement used in models of heat loss, whether for individual houses or
the whole housing stock, vary considerably (Table 4.2). The highest rates
are associated with old dwellings and with fuel combustion in individual
appliances in the absence of CH. Because it is a linear relationship,

reducing the air change rate from 4 ach to 2 ach halves the heat loss from ventilation, so the range of values quoted in Table 4.2 of 0.6-4.0 ach does represent a variation of more than sixfold. The effect of this cannot be estimated, in the absence of accurate survey data identifying the proportions of the housing stock at different levels of 'leakiness'.

In comparison with the figures quoted in Table 4.2, 0.5 ach has been the statutory requirement in Sweden for over twelve years (Olivier 1985, p.79) and heat-reclaiming mechanical ventilation systems are now obligatory. Both of these factors greatly reduce the heat loss from essential ventilation (Scandia-Hus leaflet, 1986). Against this standard, British homes are inherently more leaky and 'the method of construction and the average quality of workmanship' are contributory factors (Nevrala 1979, p.5). Therefore, draughtproofing existing doors and windows, or specifying tight windows and external doors, 'may not be sufficient to control air leakage' (ibid).

Table 4.2 Domestic ventilation rates assumed in various models or surveys

Author/Report	Ventilation rate assumed, during the heating season
Leach and Pellew	0.6 ach in winter, in most categories 2.4 ach for NCH homes using solid fuel or free standing paraffin, propane and butane fires
ETSU	1.0 ach for post-1965 dwellings, rising to 1.5 ach for pre-1919 dwellings
Penz	0.8 ach for post-1951 1.5 ach for pre-1950
BRE	1.20 for post-1918 1.70 for pre-1918
Olivier *et al.*	1.5 ach for 1939-83 2.5 ach for pre-1939
Watt Committee	3.0 ach for average, mid-19th century terraced house
NBA Tectonics	4.0 ach for unimproved, NCH, pre-1919 house with solid walls
EIK Project	not stated, assumed to be 1 ach

Sources: Leach and Pellew 1982, p.42; ETSU — Evans and Herring 1989, p.29; Penz 1983, p.86; BRE — G. Henderson 1985, Building Research Establishment, pers comm; Olivier *et al.* 1983, p.63; Watt Committee 1979, p.28; NBA Tectonics (undated), pp.28-33; EIK Project 1980.

Although ventilation rates are quoted in terms of ach, these are related to the volume of the dwelling for inclusion in heat loss measurements (0.33 x ach x volume). In order to calculate the volume, the floor area and room heights are needed, both of which are easy to measure on site, but poorly quantified on a national basis. Older dwellings have, on average, larger floor areas than recent ones (DOE 1988, p83). If both the volume and the rate of air movement in older properties are higher than for more modern dwellings, ventilation losses are strongly correlated with the age of the dwelling. Even where draughtproofing is undertaken, there could be a high rate of heat loss from ventilation in old dwellings.

The importance of ventilation rates, and the reason why they are separated out, is that they vary substantially. In addition, the level of ignorance about all the components used in calculating ventilation loss means that the rate has to be guessed, even when on site. Because of the lack of measurement methods, making an educated guess is difficult — there is no way of checking whether a house has 1.0 or 2.0 ach — so a 100 per cent error could easily be obtained. There is an urgent need for both better data and measurement methods in order to quantify

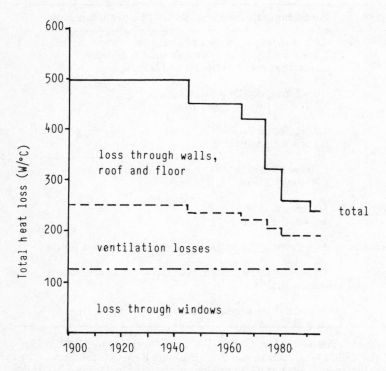

Figure 4.1 Theoretical heat loss for a mid-terrace house, by date of construction

Note: Ventilation rate of 1.6 ach in pre-1944, reducing to 0.6 ach after 1981.

Source: Tables. 4.1 and 4.2

accurately the significance of this variable, particuarly in low-income homes.

Building heat loss — COWI 1 + COWI 2

To assess the total amount of heat lost from a building, the fabric and ventilation losses have to be combined (Figure 4.1). The total heat loss halves between 1945 and 1990, though the real dividing line between 'old' and 'new' comes in 1974 — houses built since then are relatively well insulated. This figure is for a standard new house, of identical dimensions in all periods.

With improved U values for walls, floors and roofs, the proportion of heat lost from ventilation and through windows is growing. In the new standards, in this mid-terrace house, 27 per cent of the heat loss will be from air infiltration and an additional 52 per cent through the single-glazed windows. This indicates the lack of attention given to window standards in the Building Regulations and, in particular, the

> startling difference between the attention which is given in the [Building] Regulations to thermal insulation and that given to heat loss by ventilation. (*Architects' Journal* 1985, p.20)

The uncertainty over ventilation losses makes all heat loss calculations somewhat speculative, particularly at the level of models of the whole housing stock (see below).

As a result of improvements in thermal insulation, the number of dwellings built in each time period indicates the proportions of the housing stock with different levels of heat loss. Table 4.3 gives the figures, for the private and public sectors, by construction date, to tally with the effective Building Regulation dates. Within the private sector, over half

Table 4.3 Stock of permanent dwellings by date of construction start, UK 1988 (000s with % in brackets)

	Public sector*	Private sector	Total
Pre-1919	500 (2)	5,500 (24)	6,000.(26)
1919-44	1,000 (4)	3,400 (15)	4,400 (19)
1945-65	2,500 (11)	3,494 (15)	5,994 (26)
1966-74	1,400 (6)	1,850 (8)	3,250 (14)
1975-81	649 (3)	1,121 (5)	1,770 (8)
1982-88	270 (1)	1,280 (6)	1,550 (7)
Total	6,319 (27)	16,645 (73)	22,964 (100)

Note:
* includes housing associations and housing co-operatives

Main sources: DOE 1988, p.12; DOE 1989b, pp.7,28

the rented properties were built before 1919 (GHS 1987, p.125) demonstrating a considerable heat loss penalty for privately-rented accommodation.

Only 15 per cent of the domestic building stock has been built with improved standards of thermal insulation following the 1974 Building Regulation amendments. Fabric heat loss of the remaining 85 per cent of British homes is higher (for otherwise identical properties), often double the rate permitted by current regulations (from the index in Table 4.1). As the majority of modern private sector houses are probably owned by families with mortgages (and therefore not low income), and as, on average, 50 per cent of public sector tenants are poor, less than 0.5m low-income households have benefited from the improved thermal standards in the Building Regulations since 1974. Better-off households have benefited disproportionately and occupy 86 per cent of the homes with higher levels of insulation in the original building fabric.

Defective buildings

The original building fabric can deteriorate, resulting in an increased rate of heat loss. The effect of disrepair or defects on the energy efficiency of the building is unknown. No suggested method of quantification, or even discussion of the issue, has been found, beyond the expectation that there can be higher rates of air movement quoted in older buildings (Table 4.2). The deterioration occurs partly because it has had longer to accrue, partly because of the problems associated with some features of old buildings, such as sash windows. In addition, the British climate causes disrepair, because it is:

> one of the two or three most severe in the world for buildings. This is largely because the temperature passes through the freeze/thaw point more often than in almost any other climate. (Allen 1984, p.30)

The result is crumbling brickwork, stone and mortar, particularly on surfaces exposed to the rain and in the colder regions.

The 1986 English House Condition Survey found that the average cost of necessary repairs increases with the age of the dwelling, from £210 for post-1964 homes to £2,297 for pre-1900 ones (DOE 1988, p.24). In all cases, the major costs concerned the external fabric. One-third of all dwellings require repairs to the windows or walls and a quarter of all roofs (ibid, p.23). In the most serious cases (5-10 per cent of properties), the local authority can declare a property as 'unfit for human habitation' (see Glossary) and improve it compulsorily. Under the Local Government and Housing Act 1989, adequate thermal insulation is not required in a fit property, so the absence of thermal insulation does not render it 'unfit'. Other properties can lack amenities and thus present the occupier with problems in keeping warm. This is disproportionately a failure of privately-rented properties (DOE 1988, p.91):

> Ideally I would like a bathroom and an inside WC, but I'm used to my

situation. Although it's hard in the winter to go to the public baths. (An elderly private tenant quoted in Smith 1986, p.24)

Other examples of defective buildings are those with 'fundamental faults in the design and/or construction of the buildings' (NCC 1982, p.2). Often built using non-traditional methods in the 1960s and 1970s, these buildings may have such major problems, for instance condensation, that they are prematurely demolished (AMA 1984, p.54). The remaining buildings contribute to the problems faced by local authorities. In 1985, the DOE conducted a survey of the expenditure required in English local authority houses. Of the 4.6m dwellings, 3.8m (84 per cent) are in need of renovation costing an average of £4,900 to give a total expenditure of £18.8bn (DOE 1985, p.6). Of this expenditure, about £880m needed to be spent on 'minimum emergency insulation measures' in 1986 (HC 395 1990, p.vi). By 1990, urgent repairs (no improvements) were needed to 49 per cent of all local authority properties in England (DOE 1990a, Table 6).

In the previous chapter, the main characteristics of the poor were linked to economic or social status, such as being unemployed or retired, a lone parent or family with children. There is little information about households and the accommodation they occupy that is collected on a similar basis, though the 1986 English House Condition Survey found:

> Poor housing was associated with low income. Half of all households lacking amentities and one-third of those in unfit housing had net annual incomes of less than £3,000. A high proportion of dwellings in poor condition were occupied by households whose head was aged 75 or over. (DOE 1988, p.6)

Thus, there is evidence that many, and perhaps the majority of, low-income households in all sectors are living in homes that are more expensive to heat partly because of defective building fabric and that this is not a problem faced by higher income groups. As the rate at which houses are being demolished is less than 0.05 per cent of the total housing stock per annum — 11,000 out of 23m dwellings in 1988 (*Social Trends* 1990, p.128) — and because new building rates are low (Table 4.3), improvements to the homes of low-income families are dependent upon investment by government in the existing housing stock.

Opportunities for energy efficiency

The substantial repairs necessary to the housing stock have two major implications, one positive, one negative. First, any building that is defective is probably one stage away from having insulation added, because it is not usually cost effective to improve the energy efficiency of a building with poor quality fabric or limited life. Secondly, although the repairs will cost a large amount of money, they do represent an ideal opportunity to include energy efficiency improvements at relatively little extra cost. Viewing rehabilitation as an opportunity is the approach advocated by the DOE (1986d) and the Audit Commission (1986a). As the latter stated, one reason for minimizing tenants' energy bills, which 'can easily amount to over 40 per cent of rents' is to reduce rent arrears (ibid, p.51). So, it is important that:

> when property is modernised, the opportunity is taken to install insulation and draughtproofing, ensure that the heating system is efficient and that thermostats control temperature in the main living rooms. Some authorities have undertaken partial modernisation, omitting some obvious energy-saving measures in order to make the available resources stretch a little further. This may be attractive politics, in a sense that something is done to more homes, but it is not sound economics in the long run. (ibid, p.52).

However, there is a real risk that the opportunity to incorporate energy efficiency improvements in government-funded schemes will not be taken. Even where they are included, energy efficiency might not be given priority. The conflict was highlighted by the Comptroller and Auditor General:

> The Department of Energy wish to to see the resources for the Homes Insulation Scheme maintained, and indeed expanded, to include cavity wall insulation but recognised that the Department of the Environment must also take account of other priorities for housing resources. (NAO 1989, p.32)

The effect of resource constraint on energy efficiency measures is illustrated by a history of 'enveloping'. This useful initiative was developed by the City of Birmingham, primarily to improve the exterior fabric of a group of houses. Typically, the properties treated are terraced, pre-1919 and solid walled, often occupied by elderly people who are unable to maintain their properties adequately. The work is free of charge to the owners and occupiers of the building, as it is hoped that it would stimulate further investment in the interior of the dwelling. Energy efficiency improvements, such as loft insulation, cannot usually be included, even when the roof is replaced, because of the financial restriction of £11,000 a dwelling (£14,000 in London), imposed by the Secretary of State for the Environment (Circular 12/90).

The omission of energy efficiency measures from enveloping schemes is unfortunate, as the unit cost of some energy efficiency improvements is reduced through economies of scale. There are few savings where the work is highly labour intensive such as draughtproofing. Where heavy equipment is involved, for instance with cavity fill, the per dwelling cost can be at least halved as a result of greater productivity and reduced administration. The South London Consortium quotes a figure of $£2/m^2$ of wall area (excluding windows), if about 25 houses are treated together (1989, pp.14,24). This brings the cost of cavity wall insulation for a terraced house to below £100 in 1990, about the same as loft insulation in the same house.

A variety of agencies assist householders, particularly elderly people, to obtain packages of home improvements, often utilizing grant money. Many agencies would like to increase the amount of energy efficiency improvements provided, but cited lack of expertise, suitable contractors and finance as the main hurdles (Owen 1989b, p.5).

As identified in Chapter 3, over 70 per cent of low-income households live in rented accommodation. The legal situation is that the landlord, whether public or private, is responsible for repairs and for mains services, but is not required to undertake improvements, such as adding insulation. Tenants have rights to get some types of work done either at their own expense or at the landlord's, but many are unable or reluctant to exercise this right because they have no capital, or do not wish to invest it in the landlord's property, or for fear of repercussions. The large numbers of rented properties that are in disrepair demonstrate that neither these legal responsibilities nor rights are being exercised adequately, perhaps, in the private sector, because of the complex financial relationship between investment and rent levels.

The level of maintenance and repair taking place in both the local authority sector and private low-income housing is insufficient to prevent continuing deterioration. The standard of thermal insulation of the worst housing is not being improved through the fitness standard, despite the greater levels of thermal insulation required by the Building Regulations. In order to prevent increasing discrepancies within the energy efficiency of the housing stock, the minimum standard should be raised in parallel with the proposals for new houses. As this is not happening, the worst low-income homes are likely to deteriorate further, certainly relative to the homes of other households, and probably in absolute terms as well. There are many opportunities to improve the energy efficiency of the worst housing, but few of them are being taken.

Added insulation

The thermal insulation standard provided by the original fabric can be modified by subsequent alterations. The opportunities for adding insulation vary according to the original method of construction and the cost of the improvements, but alterations to an existing building are unlikely to achieve the thermal insulation standards of the present Building Regulations. In most cases this is because the cost cannot be justified in terms of increased value to the property, reduced energy expenditure or lower maintenance charges (economic issues are considered more fully in Chapter 9). Often, even partial work is not done for other reasons, such as lack of economic incentives (with landlords and tenants), finance or awareness of the benefits.

Insulation levels in the whole housing stock

'Insulation' is an umbrella term, rather like 'health'. There are about nine standard insulation measures that could be installed in a dwelling: cavity wall, solid wall, secondary glazing, double glazing, floor, hot water tank, pipes, loft, draughtproofing. The last four are commonly referred to as 'basic insulation measures'.

Added insulation is a relatively new phenomenon: insulated hot water tanks have been included in some new buildings since 1945 (Cullingworth 1966, p42), loft insulation was first required in the 1966 Building Regulations (Table 4.1) and insulation under ground floors only became mandatory in the 1990 Building Regulations.

There is a strong relationship between the presence of an insulation measure, ownership of CH and socio-economic group (Table 4.4) and loft insulation is the only energy efficiency measure likely to be found in most dwellings. The sample includes recent construction, where the insulation was automatically provided to comply with the Building Regulations. There is a close correlation between the youth of a building and the presence of an insulation measure or double glazing (DOE 1988, p.94). British homes are poorly insulated and rarely contain added insulation measures. This means that the higher rate of heat loss in the original fabric of older dwellings is not being compensated for through the addition of insulation.

Table 4.4 Types of insulation in the home and the presence of central heating, by socio-economic group, GB 1986 (% of households with measure)

	Loft insulation	Full double glazing	Cavity wall insulation	Cavity + loft + double glazing	Any type of CH
AB	87	23	23	17	95
C1	80	17	18	10	76
C2	77	13	15	6	73
D	69	9	9	1	63
E	68	7	9	1	55
Average	76	14	15	7	73

Note: socio-economic groups explained in glossary

Source: *Mintel*, March 1987 pp.68, 73

Government funded insulation programmes

Three central or local government funded programmes have aimed to increase loft insulation (private sector originally, now claimants only), draughtproofing (social security claimants) or general energy conservation measures (local authority properties). Some of the installation of the first two measures has been carried out by community insulation projects — an initiative funded by the Department of Employment. These schemes have, therefore, involved four central government departments: Energy, Environment, Employment, Health and Social Security, with the potential for inconsistency and duplicated bureaucracy. Any original expectation that the programmes could knit together has unravelled, not least because of the Conservative

Government's desire to reduce social security and local authority expenditure.

This section examines these initiatives to see which income group has received the most assistance and the extent to which income-based disparities have been increased, or decreased, as a result of government finance. Whilst the basic insulation measures are the most cost effective, several other energy efficiency improvements can give a real rate of return that is in excess of 8 per cent (the government's minimum requirement for public sector investment — discussed in Chapter 9). However, grant aid has, so far, been limited to basic measures.

Recent government policy has been to achieve value for money by ensuring that grant recipients are not 'free riders' (Macintyre 1985, p.28). The intention is that money should be given only to people who would not otherwise have undertaken the investment, or not as soon, so that the grant demonstrates 'additionality'. The DOE expresses similar views with regard to home improvements in the private sector:

> There is no need to offer help to those who would have undertaken the work anyway. (Cmnd. 9513, 1985, p.26)

Thus all insulation schemes are now targeted on the poor, particularly those in receipt of a means-tested benefit.

Loft insulation

The original aim of the Homes Insulation Act 1978 was to reduce energy consumption rapidly in private sector housing through a 66 per cent grant (Hansard 19 May 1978, col 952), as the public sector was covered by a separate programme. To encourage the private sector to take up the grant it covered both materials and labour, and has continued to do so. Various changes gradually extended the scheme and introduced a 90 per cent grant for some low-income households (NAO 1989, p.46). On 1 February 1988 the 66 per cent grant was abolished and eligibility for the 90 per cent grant extended:

> By definition, those must be the poorer people and the people most in need... We think that the number of eligible householders will increase from 4.5m to about 7m. (Hansard, House of Lords, 11 February 1987, col 645-6)

Thus, a scheme originally targeted on larger energy users is now solely for those on a low income and without capital.

The majority of loft insulation grants (66 per cent and 90 per cent) in England have gone to owner occupiers and about 8 per cent to public sector tenants; there is no information on how many have gone to private landlords or their tenants (DOE 1989b, p.25). It can be assumed that all the 90 per cent grants have gone to low-income households (Table 4.5), but the allocation of the 66 per cent grant between income groups is more problematic. The estimate of 20 per cent for low-income households is believed to be generous and implies that £93m of both grants has gone to low-income households and £152m to other households. Therefore, the loft insulation grant has provided most assistance to the more affluent households, though the balance will swing towards the poor now that they are the only eligible recipients.

Table 4.5 Government-funded capital expenditure on energy efficiency improvements, GB 1978-89

	Loft insulation grants 66%	90%	Draught proofing	LA energy conservation	Employment initiatives
TOTALS					
Properties treated	3.7m	0.6m	0.7m	2.4m	—
Expenditure	£190m	£55m	£30m	£184m	£90m
LOW-INCOME HOUSEHOLDS					
% of total	20	100	100	50	100
Properties treated	0.7m	0.6m	0.7m	1.2m	
Expenditure	£38m	£55m	£30m	£92m	£90m

Note: see text for explanation of percentages

Sources: Mainly DOE 1989b; DOE and EEO — pers comm; NAO 1989; Owen 1989a; Hansard 10 March 1987, WA col 110 and 12 Dec. 1989, WA col 535; author's estimates

The number of 90 per cent grants peaked in 1982, but has reduced subsequently, probably reflecting the declining value of the 90 per cent grant, the inability of low-income households to pay the balance (10 per cent) and inadequate publicity. Expenditure in Britain, including administrative costs, reduced from £12m in 1988/89 to £6m in 1989/90 (Hansard 9 May 1990, WA col 159; 11 May 1990, WA col 250). As these figures are for actual expenditure on 90 per cent grants only, the government's expectations of growth have not been realized. After 10 years, less than 10 per cent of all eligible households have received a 90 per cent loft insulation grant.

The recommended thickness for loft insulation has changed over the years, both when getting a grant and to comply with the Building Regulations (Table 4.1). This means that early installers of loft insulation have an inadequate amount of insulation by present standards. By 1986, only 40 per cent of lofts had 100mm or more of insulation. The remaining 10.5m homes had no or inadequate insulation in their lofts (Henderson and Shorrock 1989, p.11). Thus, many of the 1.3m (Table 4.5) grant-aided low-income households may need additional loft insulation now.

Draughtproofing

Reducing ventilation losses through draughtproofing is one of the most cost-effective ways of improving the energy efficiency of the home. However, there is no way of measuring the standards achieved by draughtproofing and the durability of materials varies considerably, so no qualitative or quantitative data on the installation and effectiveness of draughtproofing measures are available. In addition, the presence of draughtproofing might mean that only one window has been treated. By 1986, nearly three-quarters of homes with draughtproofing had done less than 40 per cent of their rooms (Henderson and Shorrock 1989, p.14). Only 33 per cent of socio-economic groups D and E have any draughtproofing in their homes (Hansard 10 May 1990, WA col 201). Householders in receipt of SB, and with less than £500 in savings, were able to claim a single payment for the cost of draughtproofing materials, but not labour costs, from the DHSS until April 1988. Since then the Department of Employment has administered the Energy Grant — misnamed because it still only covered the cost of draughtproofing materials. The grant was reduced to 90 per cent and was available if, and only if, the work was undertaken by a community insulation project. Eligibility was extended to anyone on IS, HB or family credit. Thus, within the catchment area of an insulation project, more households became eligible, though they have to find £5 towards the costs. If there is no project in the area, there is no grant, even if the householders wish to undertake the work themselves. This scheme, therefore, set the unfortunate precedent of converting a DHSS single payment into a grant dependent upon employment initiatives. Poverty and need were no longer the trigger for the payment, nor was the energy inefficiency of the dwelling.

By 1989, 700,000 properties had been draughtproofed with a grant for materials at an average cost of £43 (Table 4.5). Therefore, about one in four of low-income households with some draughtproofing had received a grant towards the cost. The rate of work by projects in 1990 is expected to have increased to about 120,000 jobs a year and double under HEES (see below).

Community insulation groups

The most direct assistance to fuel poverty sufferers, through energy efficiency improvements to their homes, has come from community insulation groups, supported by Neighbourhood Energy Action (NEA) and Energy Action Scotland (EAS). Funded though various schemes by the Department of Employment, to provide work experience for the long-term unemployed, these projects are able to provide free labour. With the 90 per cent grant for loft insulation from the DOE, grants for draughtproofing materials from the DHSS or Department of Employment and administrative costs covered by the EEO, these groups are able to provide basic insulation and advice to the disadvantaged free of charge or at a minimal cost. The 700,000 households helped since the scheme started in Durham in 1975 have received draughtproofing and occasionally loft insulation.

The cost of the labour element is difficult to assess (Table 4.5). For instance, the NAO give £23.2m as the 1986-7 cost on employee wages and the Department of Employments administrative costs (1989, p.33). No further breakdown is available. Additionally, any trainee programme is a combination of low wages and low productivity so that it is difficult to use as an exemplar.

The clients of community insulation groups mainly live in public sector housing (78 per cent) or are owner occupiers (15 per cent); only 7 per cent are in privately rented accommodation (Hutton *et al*. 1985, p.9). Thus, the majority of the expenditure on community insulation projects, £45m in 1987-8 (Hansard 9 November 1987, WA col 26), is being used to enable the previously unemployed to draughtproof local authority dwellings.

The work of community insulation groups is vital for low-income families and for the employment that it creates, but the effect on the energy efficiency of the homes of the disadvantaged is negligible in comparison to the scale of the problem. In total, about 12 per cent of the poor have received minimal insulation assistance and advice in the 15 years since 1975.

Local authority energy conservation programme

In 1978 local authorities started what was originally envisaged as a ten-year rolling programme to provide basic insulation in all public sector dwellings. Originally the money for the energy conservation programme was hypothecated, or earmarked, for this purpose within the housing investment programme. This procedure ceased in 1980 and, combined with the constraints on local authority finance and reduced rate support grant, has resulted in the decline of the local authority energy conservation programme. In 1988 work was undertaken on only 9 per cent of the number of homes treated in 1979 — 60,000 instead of 632,000 (DOE 1989b, p.25).

Under this programme, work has been carried out on about 2.4m dwellings (approximately a third of the total local authority stock) at an average cost of £40 per dwelling in 1978, rising to £137 in 1988. This includes labour costs, whether done by council employees or outside contractors (DOE 8 July 1987, pers comm). In practice, a minimum amount of work has been undertaken per dwelling, as these average costs are so low: £160 is the estimated cost in December 1985 of the materials for loft insulation, draughtproofing and cylinder cover (Hansard 1 Dec. 1986, WA col 455-6). Thus it has taken eleven years to improve marginally the insulation of a third of local authority owned dwellings: the remaining two-thirds of the stock would take another 70 years to treat at the 1988 rate.

During the period 1978-88, only about half of the residents of local authority dwellings had incomes in the lowest three deciles. So that, out of the total expenditure under this programme, only £92m is assumed to have benefited the poor (Table 4.5)

Capital expenditure on the poor

These insulation programmes have resulted in a total capital expenditure on materials and labour for low-income households of £300m over the period 1978-89. There are no comparable programmes to improve heating systems in low-income households (Chapter 5). The only other identifiable sum is through general local authority improvements, such as Estate Action. This programme is obtaining significant sums of money with 23 per cent spent on heating and insulation: £44m in 1989-90, the fifth year of operation (DOE 1990b, p4). Based on previous reports, the majority of this is probably going on heating systems. It is also uncertain how much benefits the poorest 30 per cent of households. Therefore, the £300m identified in Table 4.5 is not the total sum, but probably accurately reflects the level of government investment in — and therefore commitment to — energy efficiency improvements in low-income homes. In comparison, as shown in Chapter 3, the same poor households spent £2,600m on fuel out of state-derived income in 1988 alone. There is, therefore, a gross imbalance between government capital and revenue expenditure: the capital expenditure over 12 years is 1 per cent of a single year's revenue expenditure.

A further issue is the cost effectiveness of different methods. An accurate comparison is impossible with the data available, though community insulation projects do appear to be expensive. This may be quite justifiable as they provide advice to clients as well as providing training for employees. However, more detailed analysis is required of the costs and benefits of the different methods of financing energy efficiency improvements in low-income households, to ensure that the best value is being obtained for public money, particularly as local authority energy conservation work is being cut back, whereas community insulation projects have additional funds to work mainly on local authority properties.

In total, a generous estimate of the number of low-income households assisted by government-funded insulation programmes is:

— 0.7m with draughtproofing only,
— 1.3m with loft insulation grants and
— 1.2m with a mixture of these measures, through local authority energy conservation work.

It is assumed that poor families need the assistance of grant aid and are unlikely to undertake unfunded work, so that approximately 2.5m low-income households with basic insulation measures (loft insulation and/or draughtproofing) represent the extent to which poor homes have had insulation added.

New initiatives

Minor Works Grant

The Minor Works Grant came into operation in England and Wales in April 1990 with 100 per cent grants for low-income households in receipt of a means-tested benefit (IS, HB, Community Charge Benefit or Family Credit). Eligibility is restricted to those in privately rented accommodation or who are owner occupiers. A household can claim £1,000 up to three times, primarily for thermal insulation improvements; the elderly can obtain assistance for heating system improvements as well. The Minor Works Grant subsumes the old Homes Insulation Scheme and thus loft insulation, which was a mandatory grant. Unfortunately, the local authority now has discretion over the provision of money for the Minor Works Grant and may exercise its right to make no money available. It is not yet known what funds are being allocated.

Home Energy Efficiency Scheme (HEES)

It is proposed that from January 1991 low-income households will have a combined grant, known as HEES, for draughtproofing, loft insulation and advice. The money will include labour charges where necessary. The HEES will provide the mechanism for community insulation groups to develop into businesses, thus ending their dependence on employment initiatives and creating a universal scheme. This decoupling from the Department of Employment, the greater administrative simplicity and national coverage are all important developments. After many years of *ad hoc* policies, some rationalization is sorely needed. However, the proposal is still to provide only basic insulation measures and to require a contribution of up to £15 from claimants, who are known to be amongst the poorest in the country. Also, the suggested budget is extremely limited in comparison with the potential inherent in a national network. There have, therefore, to be serious reservations as to whether the HEES will be able to deliver the proposed service.

The legislative base for the HEES is the Social Security Act 1990. This would allow a much more extensive scheme to be developed as it permits grants to be given:

> for the purpose of improving the thermal insulation of dwellings, or ... of reducing or preventing the wastage of energy in connection with space or water heating in dwellings. (Section 15)

If the HEES can be established with a strong array of insulation projects and businesses, then the legislative and administrative structure is available for a growth of targeted assistance on low-income households. Although a substantial 'if', the HEES does represent a real hope for the fuel poor.

National heat loss characteristics

All of the above data on the original fabric, ventilation rates, added insulation and disrepair can be summarized in the total heat loss of an individual house and thus used to model the whole housing stock. The total heat loss is required in calculations of the amount of energy needed to achieve specified standards. As described already, there are particular problems about how to assess ventilation rates and to model the negative effects of disrepair. For both these, and other, reasons, the actual rate of heat loss can differ substantially from that predicted by an exercise using standard U values (e.g. a BREDEM-based computer audit — see Glossary) and needs to be cross-checked against temperatures achieved and energy consumption.

Good basic data are provided on heat loss rates from detailed measured surveys of samples of local authority housing in Birmingham (EIK Project 1980) and all housing in Cambridge (Penz 1983). There are no surveys covering a broader section of the whole housing stock. These two surveys cover just over 5,000 properties and show that within each dwelling type, on average, the worst house lost heat six times faster than the best house. If all the other factors influencing the cost of warmth (COWI 3-7) were held constant, the worst semi-detached house would cost more than 17 times the cost of heating the best flat.

The net effect of these variations is that building type is a poor indicator of likely heat loss. Because of the way that rates of heat loss are calculated, it is true that the total heat loss is proportional to the amount of external surface. However, most dwelling types span several building periods and are of varying proportions and construction details, so that the amount of external surface is only loosely correlated with the number of external surfaces. It is an over-simplification to assume that dwelling type (and thus the number of external surfaces) can be used as an accurate proxy for total heat loss. In practice, there is as wide a variation in heat loss within each dwelling type as there is between groups. This finding is of considerable importance for the targeting of assistance. For instance, Hutton et al. have recently suggested that a fuel allowance for low-income households could be paid as:

a simple, flat-rate addition to housing benefit for different types of accommodation. (1987, p.19)

Whilst administratively simple, and attractive for that reason, the evidence indicates that this proposal is unlikely to target help on the most energy inefficient homes. Many terraced houses have higher heat losses than the mean for detached houses in Cambridge (Figure 4.2). Although modern, purpose-built flats were under-represented in the sample, any additional examples could only extend the variation. The increasing mean disguises much larger variations within each dwelling type.

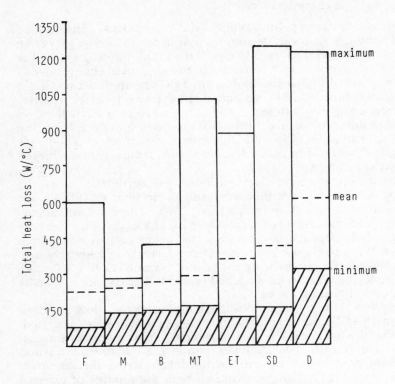

Figure 4.2 Total heat loss by dwelling type in Cambridge (W/°C)

Notes: F: flats; M: maisonettes; B: bungalows; MT: mid-terrace; ET: end of terrace; SD: semi-detached; D: detached

Source: Penz 1983, p.82

These two measured surveys and other data on housing characteristics have been used to construct models of the heat loss in the whole housing stock by three research institutes — Table 4.6 summarizes the results. The main difficulty with these data is that they are providing averages, and that these, like the means in Figure 4.2, disguise the substantial variations shown to exist within a particular group. For instance, local authority dwellings have a heat loss range of 120-620 W/°C, though a mean of 323 W/°C in Birmingham. Despite this problem, the Table provides some interesting evidence.

All three national models have been validated against total energy consumption. The ETSU model sponsored by the DEn shows a higher rate of heat loss than that estimated by BRE for the DOE, demonstrating the effect of the subjective decisions necessary because of uncertainties over ventilation rates, floor area, room heights and demand temperatures, amongst other parameters. In a paper analysing the

Table 4.6 National heat loss characteristics (W/°C)

	Local surveys		National models			
	B'ham 1977-8	Cambridge 1980-1	UK IIED 1980	GB ETSU 1982	BRE 1976	1988
Average	323	357	263	359	366	295
Tenure:						
Local authority	323	265	231		293	236
Housing Assoc.		255				
Private rented		390	283		393	315
Owner occupied		413	283		409	316
Heating:						
No central htg		339	286		361	298
Central heating		375	252		371	294
Age of dwelling:						
Pre-1900		354-599*				
1900-19		308-756*		428		374
1919-39		257-471*		380		305
1940-64		232-554*		322		271**
1965-82		71-611*		308		260**

Notes:
* averages for different dwelling types, e.g. 354 for terrace houses, 599 for detached
** slightly different age-bands

Sources: Birmingham — EIK Project 1980, p.28; Cambridge — Penz 1983, pp.82-5; ETSU — Energy Technology Support Unit, Evans and Herring 1989, p.98; IIED — Leach and Pellew 1982, p.56; BRE — Building Research Establishment, Henderson and Shorrock 1990, and pers comm with data from BREHOMES, a computer model of the housing stock (see Glossary)

changes occurring between 1976 and 1982 (not 1988 as in Table 4.6), using the BREHOMES model, half the reduction in average heat loss is estimated to have come from loft insulation:

> The remainder is shared between cavity wall insulation, draught-proofing and double glazing in about equal measure, with a smaller contribution from new dwellings. (Henderson 1986, p.61)

Total heat loss in the average house has declined at about 4 W/°C per annum for the period 1970-86, though the rate of improvement is slowing down (Henderson and Shorrock 1989, p.18).

Local authority dwellings have a lower heat loss than other tenures, substantially because the properties have been built relatively recently and include a high proportion of flats. Part of the reduction between 1976-88 comes from changes in the composition of the stock being measured as a result of council house sales: more houses have been purchased by their occupants than flats (*Social Trends* 1987, p.145). As flats, with their lower heat loss, are forming an increasing proportion of all council housing, the average has been reduced without any insulation being added. However, local authority dwellings have a lower heat loss than privately rented accommodation.

The data in Table 4.6 illustrates the confused state of knowledge on heat loss, for instance whether centrally heated (CH) or non-centrally heated (NCH) have a higher heat loss. There is, therefore, uncertainty about the significance of what has been achieved and about what needs to be done, though the increasing sophistication of BREHOMES should clarify the situation.

All the data in Table 4.6 is for heat loss from the whole house. Whilst this is important for the calculation of energy consumption, it is not the most appropriate measure for energy efficiency comparisons. For instance, energy audits usually normalize for floor area, otherwise large dwellings always appear the most energy inefficient with a high heat loss. Unfortunately, there are no comparable national data on variations in energy efficiency, particularly in comparison to people's ability to pay.

As a compromise, Table 4.7 gives income per room for households in the lower income quartile of each tenure category. This measure provides a basis for ranking the tenures and indicating the financial resources available in comparison to the space to be heated. With this analysis, not all these households are poor, nor are all the poor represented in the table. The privately rented sector contains two very different sub-groups:

Table 4.7 Relationship between household income and size of dwelling for households in the lower income quartile of each tenure, Great Britain 1987

	Rooms* per dwelling	Income (£/week)	Income per room (£/week)
Local authority tenant	4.4	48	10.90
Privately rented unfurnished	4.3	47	10.90
Housing association/co-operative	3.8	44	11.50
Outright owner	5.4	74	13.80
Privately rented furnished	3.1	67	21.30
Rented with job/business	4.9	152	30.80
Purchasing with mortgage	5.4	217	39.90
All in lower income quartile	5.0	79	15.80

Note:
* a room excludes bathrooms, WCs and small kitchens

Sources: Based on GHS 1987, pp.116,123

those renting furnished and those whose accommodation is unfurnished. Of the two, the latter have lower incomes and bigger properties, so that in terms of income per room, they are substantially poorer than the privately rented furnished category. Over the different tenures, income per room varies by nearly a factor of four. Unfortunately, a more precise comparison is not possible, so that the true extent of variations in energy efficiency in low-income households are not known.

Conclusions

There are considerable variations in the thermal efficiency and rate of heat loss of British buildings and these are probably greater than ever before. With the data inadequacies exposed in this chapter it is difficult to make comparisons between income groups, though the following is known:

— 86 per cent of the relatively well insulated homes built since 1974 are occupied by better-off families;
— the presence of added insulation is strongly correlated with the higher socio-economic groups;
— poor housing is associated with low income, elderly occupants and the rented sector (public and private).

Because of the concentration of insulation measures in the homes of higher socio-economic groups and because of the wide spread of incomes found in each dwelling type, it is assumed that within any group the buildings with the highest heat loss are occupied by low-income families. Across the whole housing stock:

— dwelling heat loss varies by a factor of at least 17;
— variations within each dwelling type are as great as those between them: there is a factor of five just within Birmingham local authority flats.

It is not possible to give proportions of the housing stock that come within different heat loss bands, nor to know the variation between bands. For instance, are the 10 per cent least efficient a factor of 5, 10 or 20 worse than the 10 per cent with the best thermal insulation? Furthermore, a true guide to energy efficiency would take into account the floor area of the property, but again the data are inadequate to indicate this range.

The present variations in energy efficiency have been created partly by government initiatives to encourage the efficient use of energy, through insulation grants and revisions to the Building Regulations. The poor have received less benefit from these policies than other households. In addition, better-off families have invested more of their own money in all types of insulation. Of the poor households it is estimated that:

— 2.5m have had loft insulation/draughtproofing/both
— 0.5m live in property built since 1974
— 3.6m live in older property, with little or no added insulation.

The main reason why low-income households in general have not been further disadvantaged by central government programmes is that those in the public sector have been compensated by the effects of the local government energy conservation programme. It is the 43 per cent of the poor who are in private sector housing (rented and owned) that have been most disadvantaged by government funded programmes. The cost of providing insulation in poor households has been £300m in total for the twelve-year period 1978-89: less than 12 per cent of the £2.6bn paid annually for fuel by the same poor families, out of state-derived income. Nearly a third of this expenditure has gone on creating employment in community insulation groups through the Department of Employment. Whilst an admirable initiative, both for the poor and the previously unemployed, the cost per insulation job completed (on local authority property) could be twice as much as it would have cost if the local authority had organized the work.

All three of the government's main insulation programmes — loft insulation, draughtproofing and local authority dwellings — have declining budgets. There has been no attempt to compensate for the deprivation caused by the cessation of HA. The failure of government to focus adequate assistance on the poor may be rectified by the new Home Energy Efficiency Scheme and Minor Works Grant, though early indications are not encouraging. Better-off families will, conversely, benefit more from the improvement to the thermal insulation standards in the 1990 Building Regulations.

5 Creating warmth

The second main determinant of the cost of a unit of warmth is the efficiency with which the heating system creates warmth. This results from both the efficiency of the heating system (COWI 3) and the cost of the fuel used (COWI 4). This continues the discussion of the factors influencing the price of a unit of warmth started in the previous chapter; the quantity of warmth required is examined in Chapter 6.

It is argued in Chapter 1, with regard to fuel-pricing policy, that government intervention is rarely appropriate or possible solely for the purposes of combating fuel poverty. However, it is important for individual households to have access to the least expensive fuel wherever possible. In this respect, COWI 3 is the last of the factors that can be influenced through capital expenditure — COWI 4 can only be altered in the access that the household has to the fuel, not in the way that the price of the fuel itself is set. It needs to be established whether the poor have systems that produce expensive or inexpensive warmth and to examine whether the relative position of low-income households has changed in this respect over the last few years.

There is a symbiotic relationship between a heating system and the fuel used — each is dependent upon the other. Very rarely can a system use more than one fuel and then the choice is limited between, say, coal and wood. There are no appliances available that can be fuelled by either gas or electricity, though most gas-fired systems need electricity for some functions such as water circulation pumps and spark ignition. Thus, the choice of a heating system in most cases results in the use of a specific fuel and substitution of the fuel depends upon capital investment on an alternative heating system or the replacement of major parts of the existing system. Capital constraints, therefore, restrict fuel substitution; switching between other commodities, for instance carrots and baked beans, does not have this connotation.

It has been established that low-income households are likely to be living in less well-insulated homes than other families in the same type of dwelling and thus the poor obtain reduced value for the warmth once produced. This disadvantage would be compensated for if the poor have access to heating systems that create inexpensive warmth, or compounded if they have systems that produce expensive warmth. If the poor are not disadvantaged in the cost of the warmth they buy, then the solution to fuel poverty is the same as for poverty — additional income.

Different approaches are needed if it can be shown that the poor have to
buy expensive warmth, for whatever reasons.

Technical efficiency — COWI 3

There can be no heat flowing from the house unless warmth is produced
within the home, from heating and other appliances, human occupancy
and solar gains (the last three are relatively minor and are dealt with in
Chapter 6). Once heat is produced, and it is warmer inside the home than
outside, there is a constant flow of heat from the building. Thus, the
production of heat itself is of benefit to the occupants only if it is produced
when they want it. A warm, empty building is not benefiting anyone.
Because of the importance of this human dimension, the technical
efficiency of a heating system is, ideally, the amount of heat produced,
where and when it is wanted, in comparison with the heat content of the
fuel consumed. This is a more specific definition of useful energy than is
used in some sources, though it is supported by a recent government
publication (Evans and Herring 1989, p.10). The interaction between
occupancy and heating levels causes further difficulties when defining
adequate warmth (Chapter 7).

There has been virtually no improvement in the thermal efficiency of
new domestic heating appliances over the last 20 years. For instance, gas
CH has improved from 64 per cent efficiency in 1970 to 66 per cent in 1986
(Henderson and Shorrock 1989, p.20). This means that replacing an old
system with a close modern equivalent will not result in a great
improvement in the efficiency with which the fuel is used, unless the
original gas CH system is more than 20 years old (50 per cent efficient) or
the replacement is a gas condensing boiler (85 per cent efficient). The
greatest improvement has occurred with individual gas fires (51-57 per
cent) and the least efficient appliances are the individual coal-effect gas
and electric fires — 15 per cent efficient if they are under an open chimney
(ie all gas and some electric). Typical efficiencies of modern heating
appliances are given in Table 5.1.

These efficiencies apply to idealized operating conditions, rather than
reality — just as the measurement of miles per gallon for car performance
is rarely achieved in practice. The efficiency of appliances might be
reduced because of:

— oversizing of the appliance in relation to the heat loss in the home
— of most significance where full CH has been installed and thus
of minimal importance in low-income homes;
— lack of controls on heat output (thermostats, timers, etc.) — again
most important in homes with full CH; limited data on ownership
of controls;
— insufficient maintenance or disrepair, perhaps from age — no
quantification of this effect has been found;

— a mismatch between the responsiveness of the heating system and the occupants' activity pattern, in both space and time. The main problems for low-income households occur with electric CH systems.

Table 5.1 Costs of delivered and useful energy, with new heating appliances, 1990

	Technical efficiency (%)	Delivered energy (£/GJ)	Useful energy (£/GJ)
INDIVIDUAL ROOM HEATERS			
Electric — radiant or convector on general tariff	100	18.4	18.4
Bottled gas (Butane) heater	92	14.9	16.1
Coal open fire	28	3.5	12.5
Paraffin stove	94	9.6	10.2
Smokeless fuel room heater with back boiler	60	4.7	7.9
Gas radiant/convector	60	4.1	6.8
CENTRAL HEATING			
Electric night storage heaters using Economy 7	90	6.8	7.5
Anthracite	70	4.1	5.9
Gas — conventional boiler	70	4.1	5.8
Oil	70	3.8	5.4
Gas — condensing boiler	85	4.1	4.8

Notes: No standing charges are included in these costs. Households using prepayment meters pay more per unit of gas or electricity. These fuel prices are based on costs in the south-east as variations occur (see text).

Source: Sutherland May 1990

The most responsive systems are individual fires burning anything but solid fuel, as the moment that one of these is switched on, or lit, it produces heat at the maximum rate: when heat is wanted it is available. Unresponsive systems, on the other hand, require time to reach peak heat output, and therefore have longer warming-up, and cooling down, periods. The proportion of the emitted heat that is of benefit to the user depends upon when that person is in the home, as well as the rate at which heat is lost through the building fabric. There are spatial as well as temporal problems with the control of some CH systems. When the heat is dispersed from a central boiler to separate radiators it is possible to turn off an individual radiator in an unused room. For electric underfloor or ceiling heating reducing the area to be heated may be impossible: to restrict heating to occupied spaces means turning off the CH and using an alternative form of more localized heat output.

There are two other problems with the technical efficiency of electric storage heating. First, the need for warmth has to be predicted up to 18 hours in advance, with the inevitable consequence that, on many occasions, more charge is absorbed than is eventually wanted, and vice versa. Estimates of the efficiency of storage systems in practice vary more than with almost all other heating methods. For instance, Chapman considered the extra charge represents 30-40 per cent more heat than is needed (1975, p.165) and Leach *et al.* give the effective efficiency of a night storage heater as around 65-75 per cent (1979, p.108). However, most comparative tables give the efficiency as about 90-95 per cent, as in Table 5.1.

Secondly, the operating efficiency of ceiling and underfloor heating depends on the way that it has been installed. With the former, an air gap around the cables prevents heat being conducted away and with the latter the thickness of cement screed both above and below is crucial. If the screed above is too thin, there is insufficient storage capacity, and if it is too thick, the system becomes a ceiling heating system for the unit or the earth below. Underfloor heating also ceases to function effectively if the floor is carpeted — the Birmingham study assessed the efficiency as 30 per cent for this reason (EIK Project 1982, para 24.3.2). These technical limitations probably explain why these systems proved so unpopular and expensive.

Thus, judging the technical efficiency of the heating system in a home is a complicated, imprecise process. No guidance has been found as to the variation that can normally be expected, other than figures for new appliances, though it is assumed that appliance efficiency does decline with age and disrepair.

Economic efficiency

There is another dimension to energy efficiency: the economic efficiency with which expenditure is converted into the energy service required. This is obtained by combining the cost of the fuel used with the technical efficiency of the system to give the cost of useful energy. The economic efficiency is, therefore, a more rigorous and comprehensive definition than technical efficiency. To these variable costs, which are directly proportional to use, should be added the fixed costs (capital, maintenance and standing charges) that have to be paid, irrespective of usage levels.

The cost of fuel — COWI 4

In Chapter 2 it is suggested that large variations in energy efficiency are a relatively new phenomenon. When most households used coal in open grates, the cost per unit of warmth was fairly uniform. The proliferation of fuels and systems, particularly from the 1950s onwards, increased the need for an awareness of actual and relative costs. A report by the BRE (undated but probably about 1962) demonstrates this new perception, with the surprised tone of the following conclusion:

> The extent of this opinion ... that floor-warming was expensive, suggests that the cost of the heating is an important factor. (Black, p.13)

It is still not universally accepted that fuel costs are important, as even the 1986 British Standard Code of Practice on Energy Efficiency in Housing has been framed in technical terms, ignoring the cost of the fuel used (though the accompanying Design Guide uses a definition of energy efficiency based on fuel costs). It may be that fuel costs are thought to be inherently unpredictable.

The real price changes in gas, electricity and solid fuels are given for each year since 1970 (Figure 2.2). The use of indices in this way provides a helpful simplification of relative price movements, but disguises the difference in absolute prices. The confusion that can result is well illustrated by comparing the data for solid fuel and electricity. The price indices for these two fuels have shown similar trends, both increasing in real terms. However, because solid fuel was less than a third the cost of electricity in delivered energy terms in 1970, it has remained at that relative value. This confirms the importance of comparing actual fuel prices when discussing heating costs.

During the whole period 1970-90, there has only been one relative price movement of significance — the lower price of gas because of the replacement of town gas with natural gas in Great Britain. This conversion peaked in 1971 and was effectively complete by 1978. The ranking of the main fuels (electricity, solid fuel, gas) in Table 5.1, both in terms of delivered energy costs (for 1970-82 see HC 401-II 1982, p.190) and useful energy costs (for 1964-80 see Leach and Pellew 1982, p.12), would have been the same throughout most of the period. Thus, fears about annual price changes altering the relative costs of the main domestic fuels are unfounded.

As an example of the lack of attention paid by government to current fuel prices in the domestic sector, the only regular source of data covering all fuels is produced by a private consultant (Sutherland 1990). The most recent are given in Table 5.1. Because these fuels are all sold in different units (for example, therms, kilowatt hours and litres) comparisons of thermal value per £ expenditure are awkward unless the delivered heat content of the various fuels is converted into similar units, as in this Table. The use of different, non-comparative units, is a main reason for public ignorance of relative fuel costs.

The cost of delivered energy is the price paid for the fuel at the meter or the doorstep, before it has been used in an appliance. The range is considerable with on-peak electricity five times more expensive than coal and four times the price of gas. There are, also, substantial variations within these average costs:

— Coal: regional averages vary by 40 per cent with the cheapest coal in Northern England at £2.50/GJ (Sutherland May 1990).

—   Electricity: each Public Electricity Supplier (old Area Board) sets its own rate based on formulae in its Licence to Supply. There has traditionally been a 15-20 per cent variation across the UK. In future this could grow, depending upon the effect of regional variations in the pricing formulae since privatization. The present off-peak tariff, Economy 7 in England and Wales, White Meter in Scotland, was introduced in October 1978 and is used by about 6 per cent of electricity consumers. A further 6 per cent are still on the earlier off-peak rates, which are at least 28 per cent more expensive than Economy 7.

—   Resale of electricity or gas: in rented accommodation, particularly private, the landlord may take responsibility for the payment of fuel bills, but charge the tenant for the amount used. The resale of gas or electricity on this basis can only be at or below a maximum resale price, set by the regulators — now the same as the credit tariff. In a survey of elderly private tenants, 11 per cent had been overcharged by their landlord and were paying more than the maximum resale price (Smith 1986, p.33)

Thus, the variations in the cost of delivered energy across the country, even at the same time, are greater than those shown in Table 5.1: the difference between the cheapest coal and the most expensive electricity is a factor of more than seven (£2.5: £18.4/GJ).

Cost of useful energy: COWI 4 ÷ COWI 3

When the cost of fuel is divided by the technical efficiency of the appliance, the result is the cost of useful energy (Table 5.1). Because the fuels with the cheaper delivered energy cost are used in less efficient appliances, the relative price differences have narrowed, though gas is the cheapest form of useful energy for most households (if it is available to them). Because the efficiency of new appliances has varied little, a family can lower their running costs only through fuel or system substitution.

The four forms of heating with the highest running costs — electric fires, bottled gas heaters, open coal fires, paraffin stoves — are all individual room heaters (IRH). The lowest running costs occur with gas, whether CH or IRH, or other forms of CH. In 1990 prices there is a range from £4.8/GJ to £18.4/GJ, a factor of over three and a half even using average costs, between the price of useful energy obtained from the various heating systems listed, when new. Regional and household variations in the cost of fuel increase this range. Heating systems that are no longer available, such as underfloor or ceiling heating, are not listed. If one of these systems is operating at 30 per cent efficiency on a White Meter tariff, the cost of useful energy is £22.5/GJ — over four times the cost of gas CH.

A household with CH is assumed to have cheaper heating than one with NCH. Although it ignores the benefits of individual gas fires, this simplification is useful as many other data sets only subdivide between these two groups.

Fixed costs

The variety of factors influencing the fixed costs, and the permutations possible, do not lend themselves to easy generalizations, particularly for low-income households in rented accommodation. Generally, fixed costs have been partially or totally ignored in assessing the economic efficiency of different heating systems. For instance, the EEO includes maintenance and standing charges, but not capital costs (EEO 1986). Five examples are given in Table 5.2, demonstrating the types of assumptions involved. These do not use sophisticated accountancy procedures like discounted cash flow, because the intention is to show the variables involved.

There is no consensus on the life of a heating system: the BRE consider that it is about fifteen years (HC 352-ii 1981, p.48), whereas the Building Services Research and Information Association assume 20 years (BSRIA 1983, p.ii). More recently the EEO are advocating the replacement of boilers that are ten years old or more (*Monergy News* 2 November 1987, p3). These may all be referring to gas CH; oil and solid fuel fired boilers deteriorate more rapidly because of corrosive by-products. The life of the system determines how the capital costs are apportioned.

The standing charge covers all uses of that fuel in the home, not just heating, and the amount to be apportioned to heating depends on what those other uses are, in that particular house. For instance, gas CH systems usually provide hot water and three-quarters of households with gas use it for cooking (Evans and Herring 1989, pp.122, 126). A fuel, such as coal, without a standing charge is usually delivered. Whilst the inconvenience of this may be a disadvantage for some people, for others it provides the opportunity (rare with fuel) to 'shop around' for the best price. In addition, it may be easier, or even necessary, to pay in advance for a delivered fuel and thus ensure that debts cannot accrue.

It is difficult to assess the costs and benefits of maintaining a heating appliance in good, clean condition and at peak performance for individual households. There are recommended procedures, such as having the gas boiler serviced once a year, but no information on how many households follow this advice, particularly by income group. Households using open fires are likely to have the chimney swept regularly, because of the risk of fire, at a cost of about £12 a year. As the effect of regular maintenance on the technical efficiency of the appliance cannot be quantified, both the costs and benefits of good maintenance are unknown for low-income groups.

Furthermore, there is uncertainty about who pays for maintenance in rented accommodation. The landlord is responsible for the provision of services so the cost of maintenance is included, theoretically, within the rent and does not form part of the tenant's fixed fuel costs. Whether this happens uniformly in practice, or whether some landlords expect the tenant to be responsible for maintenance is not known. The policy of one council is to service systems such as electric warm air and gas CH regularly, but with solid fuel central heating servicing is only carried out when repairs are necessary. Tenants are responsible for cleaning the

Table 5.2 Five examples of fixed and running costs in well-heated homes

	Gas CH (small house)	Underfloor electric (White Meter*)	Coal fires	Night storage heaters (Economy 7)	Electric fires
1. Installation cost:					
1976£	525	275		325	40
1900£			10		
2. Life: years	20	15	100	15	20
3. Annual depreciation: £	26	18	0	22	2
4. 15% interest on capital:£	79	41	0	49	6
5. Annual maintenance: £	28	0	12	0	0
6. Total system cost (3+4+5): £	133	59	12	71	8
7. % for heating	70	100	100	100	100
8. Heating system costs: £	93	59	12	71	8
9. Standing charge for heating: £ p.a.	27	13	0	13	0
10. Fixed costs for heating (8+9): £p.a.	120	72	12	84	8
11. Fixed costs payable by tenant (9+part 5): p/day in heating season	10	5	4	5	3
12. Fixed costs payable by landlord (remainder): p/day in heating season	34	21	0	26	0
13. Total fixed costs in p/day in heating season	44	26	4	31	3
14. Running costs to fixed standards: p/day	123	189	230	147	362
15. TOTAL: p/day	167	215	234	178	365

Notes: See text. The heating season is assumed to be 39 weeks (September—
May inclusive). Based on a heat loss rate of 300 W/°C (Table 4.6)
* A preserved tariff in England and Wales

Sources: Line 1: NCC 1976, p.95; Line 5: based on Sutherland 1990

chimney above an open fire and servicing systems they have installed (Lewes District Council March 1987, pers comm).

There are similar problems with the allocation of capital costs. Where the system has been installed by the landlord, the rent covers the cost. As there is no requirement that rented property should include heating, it is assumed that many tenants have, by default, to use their own appliances, with the capital costs accruing to them directly. Because of the legal restrictions on alterations to the structure, reluctance to spend money on the landlord's property and a shortage of capital, many tenants purchase cheap, individual, portable appliances, such as electric fires and paraffin stoves. Similarly, landlords can be expected to choose systems with low capital costs and minimal maintenance. Hence, the popularity of off-peak electric storage systems with local authorities in the 1960-70s. Thus, rented accommodation is likely to contain heating systems with low capital costs and high running costs, whether provided by the landlord or tenant.

Despite all of these caveats, a few generalizations can be made about the role of fixed costs. Gas-fired CH has high fixed costs, though these are still small in comparison to running costs — about 26 per cent in a well heated home. The fixed costs are a smaller proportion of the costs for other systems and less than 1 per cent for portable electric fires. Because the cost of warmth varies by a factor of at least three between different systems, the impact of the fixed costs is not sufficient to alter the economic efficiency rankings, or only to a minor extent. The situation is changed if running costs are low, because the building fabric is well insulated (DOE 1986d, p.37): in energy efficient homes, capital costs and standing charges are of greater importance. In energy inefficient homes, where running costs are substantial, the most economic fuel is gas.

Carbon dioxide emissions

At present, the main domestic fuels vary by a factor of more than four in the amount that they pollute (Table 5.3) as well as in the price to the consumer, with general tariff electricity being four times the level of gas in both cases. Economy 7 and the Scottish off-peak tariff, White Meter, are cheaper forms of electricity (Table 5.1), but result in as much pollution as the on-peak tariff. This needs a little clarification. Nuclear power stations provide part of the electricity base load — they are constantly generating electricity — and nuclear power, despite all its other problems, does not result in the emission of carbon dioxide whilst generating. However, the demand for electricity at night, from all sectors in the UK, is already greater than the amount of nuclear base load being provided, so that any new demand for off-peak electricity will have to be met from existing fossil-fuelled power stations. Hence, additional demand for off-peak electricity will result in the same amount of carbon dioxide emissions as general tariff electricity. If gas-fired power stations provide base load in future, the environmental damage will be less than with the present coal-fired capacity, at about $110 \, kg \, CO_2/GJ$, but still greater than the direct use of gas in the home. ($80kg \, CO_2/GJ$).

The use of off-peak tariffs, therefore, can only be advocated, if that use displaces existing general tariff electrical demand. No other new demand can be supported where an alternative exists, because it will result in an increase in emissions.

This evidence confirms the care needed when discussing carbon dioxide emissions to clarify whether the reference is to existing levels of emissions or to those that would be incurred by extra demand. The

Table 5.3 Carbon dioxide emissions from main domestic fuels (delivered energy)

	kg CO_2/GJ
Electricity — general tariff	231
— off-peak	231*
Gas	55
Coal	92
Oil	84

Note:
* *24-hour average*

Source: Henderson and Shorrock 1990, p.2;

figures of 231 kg CO_2/GJ for electricity is an average for the whole 24-hour period and over the whole year, with present generating capacity. The figure will increase during cold weather, when more coal-fired power stations are generating, and decrease during the summer as gas, hydro and nuclear base load form a greater proportion of supply. The level of carbon dioxide emissions from the generation of electricity in future will depend upon how much new gas-fired capacity is constructed and whether it displaces nuclear or coal-fired plant.

When the efficiency of the heating appliance is taken into account (Table 5.1), the most polluting form of heating is the open coal fire, followed by all forms of electric heating.

Ownership of heating

User preferences and options

There is little information on which features of a heating system make people prefer one type to another, assuming that they have a choice. Low running costs are nearly twice as important as any of the other factors (Atkins 1984, p.23). For households that are concerned about safety, electricity is the main choice and certainly not gas (HC 276 1984, p.148). Another factor is the ease with which householders can understand how to use the system. As some of the examples in Chapter 3 demonstrated, people experience difficulty with unfamiliar heating methods. In the 1960s and 1970s, households moving into local authority high-rise flats had rarely used CH. As a result of a lack of understanding of the principles of night storage and thermostats, people experienced a sense of bewilderment at the new technology (Hesketh 1973, p.99). Many tenants resorted to using electric fires, which they could understand, but which were even more expensive (ibid, p.4).

Other aspects of consumer choice, for instance which type of heating source — radiant, convection or conduction — is preferred by people, are either not of great importance, or just poorly researched. According to Croome and Roberts, the most pleasant environment is where the walls are about 1-2°C above the air temperature (1981, p.105), and is usually achieved as a result of convection. When radiant heaters or forced air circulation are used, the air can be warmer than the fabric, producing a sensation of stuffiness and cold at the same time (Bordass 1984, p.16).

Few low-income households have any choice over the system they use, particularly the 70 per cent living in rented accommodation. If the tenant wishes to change to a different system and has the money to spend on purchasing new equipment, the landlord may still be able to impose restrictions. For instance, many local authorities prohibit the use of paraffin stoves, as they increase the likelihood of condensation, or calor gas, because of explosion risks. Thus, many tenants, unhappy with their present method of heating, in reality may have a choice of on-peak electricity as the only alternative — the system with the highest running costs.

Users of electric night storage systems particularly dislike the lack of control they have over heat output and are offended by the unnecessary use of energy on unexpectedly warm days:

> It's wasted heat all the time. If it's a nice day we don't want it on but it's there because we didn't know it was going to be a nice day. (Tenant quoted in Sheffield 1986, section 3)

This particular study concluded that 'there is a lamentable lack of independent comprehensive information on electric heating systems on Economy 7 tariff and most particularly on consumers' assessment of them' (section 5). The same criticism appears to be valid for all heating systems. However, when given a choice between gas, electric and solid fuel CH, Brighton local authority tenants clearly preferred gas:

> In 1985, 300 properties on the North Moulsecoomb estate were refurbished and over 95% opted for gas CH. (Simmonds 1987, p.24)

The reasons for the preference are not given, though the assumption is that tenants are concerned primarily with low running costs and had perceived correctly that these are provided by gas CH. There was a rent increase to cover the additional capital investment, but this appears to have been the same for all three CH systems.

As already established, landlords are concerned with capital installation costs and maintenance charges. In the past, the gas and electricity industries have offered low connection charges in new housing estates, to reduce the capital costs for landlords, and to capture the heating market in those dwellings (Cook and Surrey 1977, p.25; HC 353 1976, p.xlix). Because the systems which cost least to install usually have high running costs, and because this often results in homes with condensation problems and thus fabric maintenance costs, the perspective of landlords may be changing. For instance, most new local authority dwellings have gas CH (Figure 2.1) and the GLC established that most of these systems are for full CH, with outlets in all, or most, rooms (GLC 1986, p24).

The legal responsibilities of landlords with regard to the provision of adequate heating systems are not clear. Although not required to provide a system, in one court case, it was accepted that extensive condensation resulted from 'the design and construction of the dwelling' and not 'that the activities and life-style of the occupiers are the cause of the problem'. It was 'irrelevant whether the building satisfied the requirements of any building control legislation at the time of construction' (Ormandy 1986, p18). Furthermore, the Court of Appeal decided, in a separate case also about condensation, that it is not an acceptable defence to claim that the tenant should spend more on heating: the landlord should have a reasonable expectation of what tenants can afford to spend (*Legal Action Bulletin* November 1985, p.156). The implication is that the dwelling should provide affordable warmth, relative to the income of the tenant.

Table 5.4 Main form of room heating, by fuel, 1961-86 (% of households)

	England and Wales			Great Britain		
	1961	1966	1972	1976	1981	1986
INDIVIDUAL HEATERS*						
Electricity				13	10	5
Gas				24	26	15
Solid fuel				16	12	6
Other				3	2	2
Total	95	84	62	56	50	28
CENTRAL HEATING						
Electricity		4	11	10	6**	6**
Gas		4	14	24	35	56
Solid fuel	4	7	9	6	6	7
Oil	1	2	4	4	3	3
Other		1	1			
Total	5	17	38	44	50	72

Notes: The table records what people say they use as the main form of heating — households with IRH often use more than one type and families with CH may use individual appliances as well, but these have been ignored in the Table. Electric CH systems include night storage heaters (even if only one is present in the dwelling) as well as off-peak and general tariff systems such as underfloor, ceiling or warm air systems — only about 1-2% of all households use the latter types (*Social Trends* 1987, p.139).
* No information on how many outlets per dwelling
** 3% night storage heaters, 3% other electric CH

Sources: 1961-72: Chesshire and Surrey 1978, p.19; 1976: NCC 1976, p.89; 1981: GHS 1981, p.48; 1986: Henderson and Shorrock 1989, p.20

Ownership patterns

Gas-fired systems provided the main source of heating in 71 per cent of GB homes in 1986 (Table 5.4), whereas electricity is used by only 11 per cent of households. Gas CH is the only system to have grown in popularity and all other systems have a decreasing or static share of the market. By 1986, 72 per cent of all households had CH and three out of four CH systems are gas-fired, so the high capital installation cost has not deterred most households. Thus, the increasing proportion of households using gas and the spread of CH are the two most dominant trends of the last 25 years.

The principal method of increasing the efficiency with which warmth is created has been through fuel substitution: only minimal improvements in energy efficiency are achieved by adjustments to the existing system or by changing the system, but keeping the fuel constant. Nationally, domestic heating efficiency has increased from 54 per cent in 1970 to 64 per cent in 1986 as a result of the growth in fuel substitution (Henderson and Shorrock 1989, p.21).

As night storage heaters are used by a minority of households, despite theoretically low running costs and a readily available fuel supply, the factors affecting heating system choice are complex and ill defined. The limited data on IRH show that gas fires retain their popularity, with most of the decline being in the use of solid fuel or electricity. However, 5 per cent of all households (1m families in the UK in 1990) use on-peak electricity — the most expensive form of useful energy.

The ownership of both CH and different types of CH varies across the tenure groups. By 1986, the range of CH ownership varied from:

— 37 per cent of the privately rented unfurnished households;
— 48 per cent of privately rented furnished households;
— 63 per cent of local authority dwellings;
— 73 per cent of housing association/co-operatives
— 72 per cent of outright owners;
— 84 per cent of households with a mortgage (GHS 1987, p.121).

Not only are tenants less likely to have CH than owner occupiers, but their CH system is more likely to be non-gas. Privately renting tenants in unfurnished property are about half as likely to have CH as public sector tenants. There is a similar distribution across income groups (Figure 5.1): the better-off families have bought themselves the heating system that provides the cheaper warmth.

Gas users

In 1976, about 65 per cent of households in each income group were gas consumers (Figure 5.2), though not necessarily using gas for heating. By 1986 there was a range from 64 per cent to 84 per cent across the deciles: the rich are now more likely to be connected to the gas supply than the poor. The proportion of all households connected to the gas supply has risen from 65 per cent in 1976 to 74 per cent in 1986, an extra 300,000 each

year. Approximately 85 per cent of dwellings are within the statutory
minimum distance (25m) of a gas main (Mongar 1986, pers comm), so that
about 11 per cent of eligible households have not exercised their option to

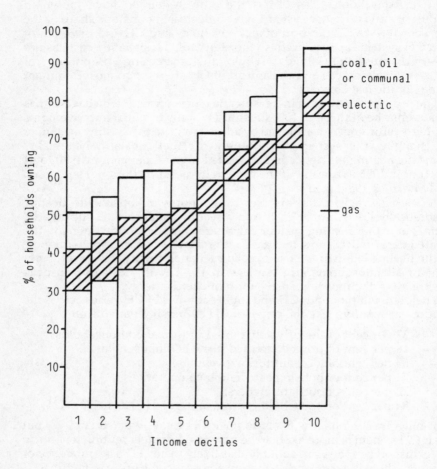

Figure 5.1 Central heating ownership by income, UK 1986

Source: Hutton and Hardman 1990, Table 7.2

be connected. Some tower blocks could not be converted to gas use,
because of the risk of catastrophic explosions, as demonstrated by Ronan
Point in 1968. The proportion of households connected has a clear spatial
distribution — over 80 per cent of homes in the Midlands but less than 60
per cent in Scotland use gas (Figure 6.1). However, there are
opportunities for both the more extensive use of gas for heating amongst
all households and more intensive use amongst present consumers.

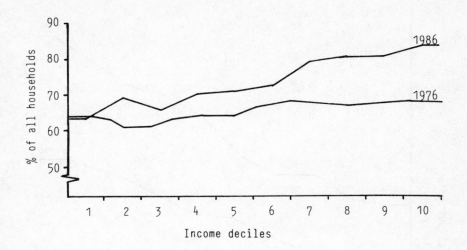

Figure 5.2 Proportion of households using gas, by income, UK 1976-86

Sources: 1976 — DEn 1978a, p.9; 1986 — Hutton and Hardman 1990, Table 3.1

Of the households connected to gas supplies, the proportion that have gas CH is shown in Figure 5.3. There has been an increase in ownership in all income groups, but this has been greatest for high-income gas users. This difference is extended by the greater likelihood of being a gas user amongst better-off households, as demonstrated in Figure 5.2. By 1986, 90 per cent of those in the top decile who were connected to the gas supply had gas CH.

Electric central heating users

Over 99 per cent of all households are connected to the electricity grid and Figure 5.4 shows the proportions with electric CH. The trend here is the reverse of that demonstrated with gas CH. In 1976, there was a relatively even distribution of ownership of electric CH across all income groups, whereas by 1986 there was a strong bias with better-off households much less likely to own electric CH than poorer families. The decline of electric CH amongst better-off households with sufficient capital to purchase alternatives demonstrates their dissatisfaction and the problems experienced with overnight charging levels and control of the heat output. Research for the Electricity Consumers' Council concluded that households:

> with non-electric CH choose it, whilst those with electric CH have it not through choice but because of where they live. (MAS 1979, p.8)

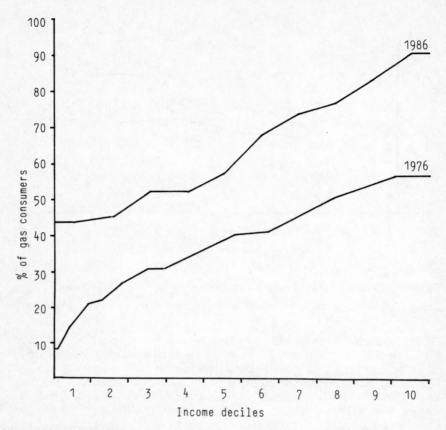

Figure 5.3 Proportion of gas users with gas CH, by income, UK 1976-86

Sources: 1976 — DEn 1978a, p.9; 1986 — Hutton and Hardman 1990, Table 3.1

Recently there has been an increase in the numbers of new local authority homes being built with electric CH (Figure 2.1) and in general ownership (Hutton and Hardman 1990, Table 7.1). This may demonstrate greater consumer satisfaction with the design of new night storage heaters, but it is increasing the problem of carbon dioxide emissions.

In 1986, 6 per cent of all homes had electric central heating, about half of these having night storage heaters (Table 5.4). A further 5 per cent of all homes use electricity for heating in non-CH systems, primarily fixed radiant fires, with portable electric fires being the next most popular (DOE 1988, Table A4.6). In 1990 in the UK this would imply:

homes with NSH	0.7m
other electric CH	0.7
homes using individual electric fires	1.1
TOTAL using electricity for space heating	2.5m

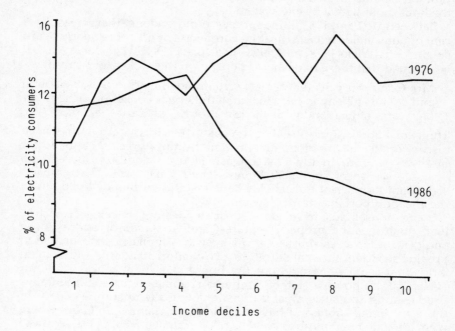

Figure 5.4 Proportion of households with electric CH, UK 1976-86

Sources: 1976 — DEn 1978a, p.9; 1986 — Hutton and Hardman 1990, Table 3.1

At least 11 per cent of all UK households depend upon electricity for space heating, almost equally divided between unrestricted and off-peak users. The distribution of electric central heating by income group demonstrates a clear bias towards the presence of electric central heating in low-income homes: 12.5 per cent of the poorest 30 per cent had electric CH, whereas only 7.9 per cent of better-off families in the UK in 1986 (Figure 5.4). Individual electric fires are probably even more concentrated in low-income homes.

Defective heating

Whether or not they are purchasing expensive warmth, some households have systems with insufficient output to heat the house adequately. Defective heating is indicative of a shortage of capital expenditure and is separate from evidence of a shortage of income. Systems may be unused for other reasons, such as the inability of the family to control or understand it, and being unable to move coal in old age. Many of these problems require capital expenditure if they are to be overcome to provide the household with a usable system.

Between 1972 and 1981, three surveys found evidence that at least 1 per cent of households have no heating equipment or do not regularly heat a single room (Wicks 1978, p.82; NCC 1976, p.100; DOE 1983, Table A.1)). It is assumed that these households depend on the cooker for warmth:

> Mrs C lives in privately rented accommodation. She is housebound and spends all day in her kitchen with her gas cooker on. This is her only form of heat and is all she can afford. (Durward 1981, p.35)

The explanation cannot be that the gas or electricity supply has been disconnected, as this does not prevent heating with other fuels, for instance paraffin. In the 1986 English House Condition Survey, 'no heating was seen' in 59,000 homes (DOE 1988, p.93). Therefore, a significant number of households have insufficient money to buy even the cheapest portable heating system.

Some households have an inadequate heating system. To define inadequate heating properly requires defined standards and a house audit, in order to establish that the available appliances are unable to provide sufficient thermal output to give comfort standards, either in an individual room or throughout the house. The new fitness standard states that, if a house is suitable for human habitation, it should contain a fixed heating appliance capable of maintaining a temperature of 18°C in the main living room and have facilities for obtaining 16°C elsewhere when outside temperatures are -1°C (DOE Circular 6/90, p.49). By these criteria, in 1986 over 1.7m dwellings (9 per cent of English homes) were considered inadequate, in all ages of property; over 1m of these inadequate heating systems are in rented properties (DOE 1988, pp.92, 94). Thus, many families are cold because they have not got adequate heating appliances.

The presence of a heating system that is deemed to be technically adequate does not mean that it is acceptable to the users. Many of the innovatory electric heating systems introduced in the 1960s were found subsequently to be unacceptable. The extent to which these, and other, systems are present, but unused, indicates the scale of dissatisfaction: a survey of 383 local authority dwellings in Gateshead found 96 per cent with CH (mainly electric), but only 57 per cent used it as their main form of room heating (Byrne et al. 1986, p.81). In some cases, the system may have been designed to provide background heating only, so that an alternative method of heating the main living room was intended. The likelihood is, however, that these electric heating systems are causing increasing problems. By 1985, there were over 500 estates where claimants were eligible for the ERHA, because the system was expensive to use; 95 per cent of these estates used electric heating (Sheldrick 1986, p.13). Since April 1988, these tenants have lost up to £8.80 per week and are undoubtedly finding it even more difficult to use the original electric heating system. These unused CH systems represent a waste of resources if they are prematurely retired, due to user dissatisfaction. Depending on what fuel is used as a replacement, there could be a further misallocation of resources if the householder is using an alternative that is less energy efficient.

The range of problems experienced by local authority tenants are summed up in the following rather cryptic results from a representative sample of English dwellings:

Rather more than a third of dwellings (37%) did not have any form of central heating; and of those dwellings with individual central heating, 13% of systems were over 25 years old. About half the households questioned by surveyors (49%) said that they had serious heating problems. The main problems were lack of heat generally (56% of problems), not enough in some rooms (29%), expense (28%), draughts (16%) and faults in the system (13%). 60% of households supplemented their fixed heating, mainly with the use of portable electric fires. Houses built before 1964 gave rise to the most complaints and ducted air and underfloor heating appeared to be the least popular.(DOE 1990a, p.6)

In summary, using figures grossed up to UK 1990 levels, the most disadvantaged households are those:

— using general tariff electricity   5 per cent or 1.1m
— with problematic electric CH systems, such as
   underfloor or ceiling heating, which they are
   using, (others have, but do not use, one of these
   systems) at least   3 per cent or 0.7m
— using open coal fires   4 per cent or 1.0m
— with an inadequate heating system   9 per cent or 2.0m

There will be some overlap between the last category and the other three, but a minimum of 2.8m families can already be identified as having particularly expensive or inadequate heating. The majority, and probably all, of these households are low income, many of them in rented accommodation.

Government policies on useful energy

Domestic heating lies on the boundary between the responsibilities of the DEn and the DOE: the DEn is concerned with the efficient use of energy and the DOE, more than any other department, with heating systems. Although there is a potential for overlap between the two departments, it appears rather that there is a policy void between them. It is still not clear whether the sparsity of policies on the efficient use of energy in domestic heating systems is a conscious or unconscious decision by different governments, though the frequent references to poor data in this chapter confirms that there is little interest in the issue.

Policies on energy use

There has been less involvement by government in this aspect of energy efficiency than in the problems associated with heat retention. This is

surprising because all administrations since April 1976 have emphasized
the importance of fuel-pricing policies in securing investment by
consumers in energy efficiency improvements that are cost effective
(Priddle 1982, pp.2- 3). Changes in the price of a single fuel are primarily
of importance to existing consumers and are most effective at stimulating
insulation measures in their homes (providing they have the capital and
legal right).

When a new heating system is to be purchased, it is relative fuel costs
that are of importance. Thus, intra-fuel comparisons are needed to
maximize energy efficiency through the installation of new heating
systems. The need to assist domestic consumers to obtain information on
relative costs was recognized during the period 1967-73, as Cook and
Surrey state:

> With the boom in central heating, the need for a fuel advisory service to
> help the consumer to choose the best fuel for his purpose was widely
> urged. Much effort was being wasted by the fuel industries in
> competitive advertising and by consumers in trying to compare and
> check cost estimates. The proposal came to nought partly because
> there were genuine difficulties in making sound generalisations about
> the relative efficiency of different heating systems for different
> buildings and living habits, and partly because of the inherent
> difficulty of forecasting future relative prices and the consequences if
> official advice turned out to be wrong. (1977, p.25)

Similarly, the Advisory Council on Energy Conservation believed that:

> many consumers with older heating systems which they are willing to
> change need better guidance from Government and the fuel industries
> on how to exploit this opportunity to reduce fuel usage and therefore
> their system's running costs. (DEn 1982b, p.4)

Periodically, since 1979, the DEn have published guides to comparative
heating costs, which are primarily designed to demonstrate the benefits
of adding insulation. The complexity of these — the most recent set has
30-page booklets for each of the four main fuels (EEO 1986) — means that
they are unlikely to provide much assistance to consumers, particularly as
neither the gas nor electricity showrooms are distributing them. Simpler
methods of conveying relative fuel prices, for instance small leaflets or
advertisements, are not used.

The Director of the EEO told the Select Committee on Energy, in
response to a question about policies to restrain the demand for some
fuels more than others:

> I do not really see that as being something we would work particularly
> closely on. (HC 87 1985, p.41)

Although slightly obtuse, the question and answer demonstrate that the
EEO is not working on fuel substitution policies for the domestic sector.
Only by identifying which fuels provide expensive warmth and actively
discouraging their use, can the government assist households in all

income groups and tenures to obtain cheaper heating. There are no policies to ensure that low-income consumers have the most energy efficient heating systems.

This reticence to develop policies to assist fuel substitution in low-income homes results partly from the DEn's traditional partnership with the supply industries. Even with the privatization of the electricity and gas industries, the DEn finds a consumer-orientated approach difficult to acquire. There is a precedent for the grant aiding of fuel substitution by the DEn: between 1981 and 1987 over £50m was allocated for 25 per cent grants for the conversion of oil and (later) gas-fired plant to coal in the industrial sector (DEn 1982b, p.32).

A further precedent has been set by the MOHLG (now the DOE) as the Clean Air Act 1956 entitles householders to a grant for at least 70 per cent of the cost of installing an approved alternative. The local authority is reimbursed for about half the grant. By 1983-4, the total cost of this scheme to central and local government was £285m, the equivalent of just over £10m p.a.; the householders' contribution has been £122m (DOE March 1987, pers comm). Local authorities have discretion in defining the alternatives that are funded, and some allow non-solid fuel systems to be installed, for instance gas fires. As this is a pollution control measure, there is no focus on improved energy efficiency, so that a household may just change to using smokeless fuel in an open grate.

The substitution of alternative fuels for oil, gas and coal has been grant-aided, so there can be no objection in principle to policies to grant aid fuel substitution in low-income households. However, this has not occurred, although for many households, for instance the one million families using general tariff electricity as the main form of room heating, fuel substitution would be the single most effective energy efficient improvement. The government has no policies aimed clearly at achieving the efficient use of energy through appropriate fuel substitution policies in the domestic sector, at least through the DEn or EEO.

The rhetoric on the need for improved energy efficiency to reduce global warming has not yet resulted in any policies to assist the reduction in carbon dioxide emissions by low-income households.

Policies on heating systems

As stated in Chapter 2, British housing and environmental health standards have never placed great emphasis on heating. The DOE has clear responsibility for building standards, but has an ambiguous role towards heating systems. For instance, the Homes Insulation Act 1978 and the Circular on Energy Conservation in Public Sector Dwellings were specifically limited to insulation measures. Between 1969-80, heating had to be provided in both types of public sector housing, at least to the minimum Parker Morris standard, though the DOE did not provide explicit guidance about which heating systems should be installed and only the briefest of references can be found to running costs.

Because of problems with heating and condensation in local authority homes, a working party was set up by the DOE in 1977. The third of their reports, Domestic Energy Notes 3, known as DEN 3 (1978), advised local authorities and housing associations on how to deal with problems in existing electrically-heated dwellings. DEN 3 took a holistic approach by considering the thermal characteristics of the building fabric together with the heating system. This was an innovative approach, particularly required with systems using the thermal storage of the home, as well as that inside the appliance, for efficient functioning. The original mismatch of heating system and insulation standards had been caused substantially by the DOE's methods of financing housing and requiring low capital installation costs, so that the DOE were the appropriate source of advice on improvements. Thus, the DOE became involved, through this working party, in the problems of unsatisfactory heating systems and, therefore, in fuel substitution, albeit negatively:

> replacement of the system may not involve a change of fuel. (DEN 3 1978, p.7)

Interestingly, the DOE have not kept statistics on the numbers of dwellings requiring conversion under DEN 3, or subsequently: they are not required to monitor the extent to which their lack of guidance has resulted in a poor return on public funds.

As a result of the DOE's lack of concern with running costs, and subsequent tenants' problems, the DHSS give a special HA to SB claimants living on estates with heating systems that are expensive to run — the ERHA. Of the first 500 estates designated for the ERHA, 95 per cent used electricity (Sheldrick 1986, p.13). Recipients of the ERHA were one of the groups hardest hit by the abolition of HA in April 1988, losing up to £8.80 per week. Despite the change in DSS support, no new initiatives have been announced by the DOE to improve estates known to have expensive heating systems.

The 1990 fitness standard for England and Wales requires 'adequate provision for heating', for the first time, if a dwelling is to be deemed fit for human habitation (DOE circular 6/90). The equivalent Scottish Tolerable Standard has included satisfactory provision for heating for some time. The British Standards Institution Code of Practice for Energy Efficiency in Housing (BSI 1986) and the Building Regulations encourage, but do not require, the use of energy targets for buildings. This will highlight the importance of the heating system in the energy efficiency of the dwelling. At the moment, the Building Regulations do not include any requirements with regard to the provision of a heating system. There are stipulations about 'heat producing appliances and their structural requirements', if a system is installed, but (surprisingly) there is no legal obligation actually to install a system. Official recognition, through energy targets, that the heating system and fabric interact in determining the energy efficiency of the dwelling is a major achievement.

There have been no identifiable central government policies to assist the energy efficiency of low-income homes through fuel substitution or improvements to heating systems. Central and local government have been spending £10m p.a. on air pollution control in all income groups, through the Clean Air Act 1956. Local authorities have spent money on replacements to electric heating systems, but the cost cannot be identified. Thus, it is not possible to extend the capital expenditure analysis given in Table 4.5. Government-funded programmes to accelerate the rate of change towards more energy efficient heating systems amongst low-income households would assist fuel poverty sufferers, but are not being established.

It is very difficult to bring together the data in this and the last chapter to give the cumulative impact of COWI factors 1-4 on low-income households: about 3.6m families live in pre-1974 property with little or no added insulation and at least 2.8m households have inadequate or expensive heating. The overlap between these groups cannot be established. In an attempt to quantify the likely effect of multiple disadvantage, Byrd carried out an analysis of the 5,000 properties in the Birmingham EIK data and assessed that the 10 per cent least energy efficient houses are five times more expensive to heat per week than the 10 per cent most energy efficient (Byrd 1986, p.24). This is the only statement to quantify the effect of variations in the cost of warmth in relation to proportions in a sample. Multiple benefit and, in comparison, multiple deprivation are indicated in Table 4.4: the higher socio-economic groups have both the greatest number of insulation measures and the highest proportion of CH ownership. For policy analysis, particularly in relation to benefit levels, the range is needed with respect to family types. The Treasury stated (Chapter 2) that HB is needed because housing costs, for similar accommodation, can vary by a factor of two. Despite the inconclusive nature of the data available, there is a much greater range in heating costs.

Conclusions

There is a threefold variation in the cost of a unit of warmth from a new appliance and the systems with the greatest running costs are disproportionately concentrated in low-income homes. The impact of age, disrepair and poor control increases the costs, but to an unknown extent, so there are no indicators of the range to be found in the present stock of heating appliances. At least 2.8m households (12 per cent) have particularly expensive or completely inadequate heating systems and it seems reasonable to assume that these are virtually all low-income families. Separately, with an unknown overlap, over 500 local authority estates were designated as expensive to heat — these should be a priority for intervention, now that the tenants have lost their HA. Thus, a minimum estimate is that at least a third of the poor require capital expenditure on their inadequate or energy inefficient heating systems.

The greatest improvements in energy efficiency come from fuel substitution: the reduction in the cost of keeping warm that results from changing on-peak electricity to a gas-fired system is greater than the effect of insulating a pre-1919 building to current Building Regulation standards. The government has shown little recognition of the importance of this aspect of energy efficiency and no capital expenditure programme of significance has been identified. Because fuel substitution and improved efficiency have been concentrated in the better-off households, the slower rates of acquisition and turnover amongst low-income or renting households penalizes them. The poor are being penalized by the lack of capital investment in efficient heating systems for their homes.

Much of the problem of inefficient heating systems has been created by the lack of government standards: no regulation or planning law requires a heating system to be installed in a dwelling. A fixed heating system in the living room is part of the new fitness standard, but this is still subject to local authority discretion as to whether it really is necessary.

The cost of a unit of warmth in rented accommodation is greater than that of the owner occupied sectors, if only because of the relative lack of CH. Rented accommodation with CH is disproportionately likely to have an electric system, despite evidence of tenant dissatisfaction with this method. If tenants are given a choice of heating system, to be paid for by the landlord, they choose gas CH, demonstrating the same preference as mortgagers and better-off families. Thus, it is the division of interests and concern between landlord and tenant, rather than poor information amongst tenants, that results in energy inefficient systems predominating in rented homes. The problem seems to be endemic to the rented sector and is not confined to privately rented accommodation.

The cumulative effect of COWI factors 1-4 cannot be estimated for low-income households. It is, for instance, impossible to estimate the variation in the cost of warmth between the 10 per cent most and least energy efficient low-income homes across the whole country, though a factor of five has been found in local authority stock. The effects of multiple energy inefficiency cannot be quantified. The probability is that low levels of capital investment in the building fabric and in the heating system occur in the same households, and vice versa. Although there is a frustrating lack of good data, the emphasis on capital stocks appears justified: low levels of capital investment in the energy efficiency of low-income homes mean that the poor have to buy the most expensive warmth. As the rates of improvement on all aspects of energy efficiency are greatest amongst better-off households, the poor are disadvantaged and increasingly so: capital investment in the house and heating system are needed to combat worsening fuel poverty.

Per unit, general tariff electricity is four times as expensive as gas and produces four times the amount of carbon dioxide. Thus, policies to reduce the cost of keeping warm and policies to reduce emissions of greenhouse gases go hand in hand. Whilst no new government initiatives have yet been announced, the growing concern about global warming may give a boost to measures that also combat fuel poverty through fuel substitution.

6 Warmth needed

The fuel poor are suffering from cold homes. Whilst everyone understands the words 'warm' and 'cold', they are difficult to quantify and thus to determine who has or has not an adequate level of warmth. This chapter tackles the definition, in order to assist in both the interpretation of temperature surveys (Chapter 7) and the formulation of policies to ensure that the poor have the choice to be warm. This chapter is, therefore, focused on people's expectations and needs: how most people would define a warm home and what levels ensure good health.

The popular view is that people's physiology (or psychology) varies considerably so they need different levels of warmth for comfort. This rationale is used to explain variations in measured temperatures in the home in terms of personal preferences. The validity of this view is assessed with reference to work in controlled environments, in the absence of economic constraints. The evidence on the temperatures required for comfort and health define the internal demand temperature (COWI 6). The amount of heating required depends upon the difference between this and the external temperature (COWI 5) when an allowance is made for the gains from other incidental sources.

The level of warmth required is a function of space and time (COWI 7) as well as temperature. Most emphasis has been placed on defining temperatures — the other parameters are more difficult to evaluate. For instance, with regard to the number of rooms to be heated, what are present social expectations and minimum acceptable standards? Is one warm room sufficient if it ensures survival? Should the elderly and other vulnerable groups be able to heat bedrooms at night in order to prevent respiratory disorders? Some of these problems are confronted in this chapter and more in Chapter 10, but many of the issues require extensive discussion and the proposals made here are only a stage in that debate.

The previous two chapters have demonstrated the substantial variations in energy efficiency that occur between houses and, thus, in the cost of a unit of warmth. This combines with the data in this chapter on the quantity of warmth needed by low-income households to give the total cost of keeping warm. The hours and degrees of warmth needed depend on factors that are beyond the influence of direct government policy: the size of the family and the dwelling they occupy. Indirectly, policies may have an influence, mainly at the margin — for example, the number of people at home during the day is affected by the level of unemployment. Otherwise, the direct impact of these factors on the individual household can only be ameliorated through income support.

Internal temperature — COWI 6

There are three objectives to be achieved in heating a home:

— the comfort of the occupants;
— the health of the occupants;
— proper maintenance of the building and the prevention of condensation.

Each of these is considered and then compared with recommended design standards (for instance for heating system performance), in order to define the temperatures in a heating standard.

Activity levels and comfort

Thermal comfort has been defined as 'that condition of mind which expresses satisfaction with the thermal environment' (Fanger 1973, p.3). Baillie *et al*. make the useful distinction between comfort, as defined by this thermal comfort theory, and the value placed upon that comfort by an individual (1986, p.308). The former is physiologically determined, whereas the latter reflects economic constraints as well as social-psychological attitudes (for instance, mining communities are reputed to have warm homes, even when access to concessionary coal has ceased). The economic value that is placed on warmth is discussed in Chapter 9 and the influence of economic constraint is considered in the next chapter. The temperature that provides comfort conditions depends on the individual's activity and clothing.

Activities in the home can be grouped into four categories by the metabolic rate, or energy expenditure, that occurs: at greater levels of physical exertion, cooler temperatures are required.

sleeping — the flow of heat from the body is at a rate of 40 Watts
    per square metre of body area ($40W/m^2$);
sitting — metabolic rate between $50\text{-}60W/m^2$;
lightly active — e.g. meal preparation, child care, $65\text{-}90W/m^2$;
moderately active — e.g. sweeping a floor, bedmaking, $90\text{-}130W/m^2$.

More strenuous tasks rarely take place in the home, or only last for a short time (BRE 1979, p.1).

It is known that these average, whole body, metabolic rates vary slightly from person to person. For instance, there is a gradual reduction in the energy expenditure of a resting person during adult life of about 15 per cent between the ages of 25 and 80, and with increasing age, each activity is undertaken less energetically (Humphreys 1976, p.177). For both these reasons, the mean metabolic rate is lower in older people. However, Fanger established that when activity, clothing levels and other experimental conditions are standardized:

elderly people do not appear to prefer different thermal environments to younger people. The lower metabolism in elderly people seems to be compensated for by a lower evaporative loss. (Fanger 1973, p.8)

By trapping warm air around the body, clothing provides insulation. This is measured in 'clos', with 1 clo equal to a resistance of 0.155 m² °C/W (Humphreys 1976, p.177) — the inverse of U values for building components, so the larger the number of clos, the greater the level of insulation provided. Normal winter wear of skirt and jumper, trousers and jumper, three-piece suit or warm dress provides from 0.8-1.2 clo (BRE 1979, p2). Women tend to wear clothing with lower insulation levels than men, as trousers provide better insulation than skirts, because they cover the legs. Also, as Hunt and Gidman established, there is a general increase in the insulation value of clothing with the age of the wearer: people under 25 years wear an average of 0.73 clo whereas elderly women and men wear 0.90 and 0.97 clo, respectively. The lower insulation values of clothes worn by younger people probably reflect the influence of synthetic fabrics and fashion. Women in homes with CH wear cooler clothing than women in NCH homes (Hunt and Gidman 1982, p117). Thus, there is little opportunity for elderly people, particularly those without CH, to obtain additional warmth from extra clothes, rather than from heating.

Through tests in the controlled environment of an enclosed room (a climate chamber), the temperature that people find comfortable can be established for a range of clothing and activity levels (Table 6.1). Whether the air in a room feels warm or not is a subjective judgement, though 95 per cent of people find the same temperature comfortable, if they all have the same activity level and clothing (BRE 1979, p.2), and all would, therefore, be free from real discomfort (CIBS 1978, A1-3).

To establish the variation that occurs in activity patterns amongst the general population, an analysis has been carried out of data from a time-budget survey undertaken in the winter of 1983-4 for the Economic and Social Research Council. People, in all income groups, completed diaries to indicate how they spent their time — as reported fully in Gershuny et al. (1986). The population was divided into sub-groups, in

Table 6.1 Comfort temperatures by activity and clothing level (°C)

	0.5 clo	0.75 clo	1.0 clo
sleeping (40W/m²)	29	28	27
sitting (55W/m²)	26	24	23
lightly active (78W/m²)	21	19	17
moderately active (110W/m²)	18*	16*	13*

Note: * estimates subject to quite large possible errors

Source: Humphreys 1976, p.180

order to make comparisons between those in work and those at home
(Boardman 1985). Although subsets of the population spend different
amounts of time in the home (see below), the main population groups
distribute their time fairly consistently between different activities (ibid,
p8). With the exception of women with young children, during the day:

56-75 per cent of the time is spent sitting,
15-31 per cent lightly active and
6-19 per cent moderately active.

Using the metabolic rates for these activities (as described above) and
assuming a clothing level of 0.82 clo — the weighted average for men and
women of all ages — the mean metabolic rate for each of the main groups
in the population is obtained (Table 6.2). These fall within the fairly
narrow band of 64-74W/m^2 for each hour spent in the home and awake.
The resultant comfort temperature for each of the groups from the time-
budget survey can be established showing that all groups would
experience comfort at similar temperatures, because of their similar
activity patterns. The mean comfort temperature is 20.8°C (rounded up to
21°C for simplicity), during the hours that the home is occupied in the
day, for all groups in the population. This remarkable finding is robust, as
it is true for all the main tenure sub-groups, and considerably simplifies
the debate about comfort temperatures.

The comfort temperature of 21°C is supported by a survey in an office
where the occupants' preferred temperature was 22°C, throughout the
year, at the slightly higher mean metabolic rate of 80W/m^2, but lower clo
value. To ensure that at least 90 per cent of the workers are comfortable,
the range would have to be within 20-23°C (Fishman and Pimbert 1982,

Table 6.2 Temperature required to provide comfort for people when in the
home and awake on an average day

	Hours	Metabolic rate (W/m^2)	Temperature (°C)	% of diaries
Full-time employed	7.4	67	21.0	38
Part-time employed	9.6	70	20.3	16
Students, dependent children & others	8.6	64	21.7	5
Unemployed	11.2	66	21.2	6
Retired, sick and disabled,	13.1	67	21.0	16
Housewives with children under 5	13.0	74	19.4	7
Housewives (without young children)	13.5	71	20.1	13
Mean	10.1	68	20.8	

Source: Boardman 1985

p.109), because dissatisfaction rapidly occurs when the temperature departs from the comfort zone (ibid, p.112). This result is important because it is derived in a natural environment, rather than during experiments, unrestricted by concerns about paying the fuel bills. This is not to deny the importance of economic constraints, but the aim here is to establish what people are probably hoping to achieve, and may in fact attain in due course. Wynn and Wynn report that in Sweden a temperature of 20°C is considered an appropriate minimum for young children (1979, p.210).

At night, the assessment of comfortable temperatures is complicated by the wide range of bedclothes used, and thus insulation values: three blankets, sheet and nightwear provide about 3.5 clo, which is sufficient when the bedroom temperature is in the range 13-17°C (Humphreys 1978, p.703). There is little information on an appropriate temperature for health and comfort at night, for different groups of the population:

> At the moment one is left to guess what the practical lower limit for bedroom temperatures would be. Temperatures of much below 15°C would probably prove unsuitable (Humphreys 1976, p.178),

and would fail to provide appropriate conditions for dressing, reading in bed, evaporating sweat from bedclothes or preventing hypothermia and respiratory diseases in the very young and very old (ibid, p.178)

Health

Establishing the minimum temperatures needed to maintain health is difficult, because of ethical limitations on experiments — some evidence is given in Chapter 7. However, as Collins states, with reference to the elderly:

> decreasing indoor temperature below the comfort zone progressively influences the respiratory, cardiovascular and thermoregulatory systems and consequently the maintenance of health. (1986, p.3)

He identifies the following temperature-linked health bands:

> 18-24°C — the comfort zone, no risk to sedentary healthy people;
> below 16°C — increasing risk of respiratory disorders
> below 12°C — cardiovascular strain
> below 6°C — failing thermoregulation and risk of hypothermia, after two hours' exposure.

Thus, the risk of morbidity increases as the temperature falls and the period of exposure lengthens. This principle holds true for the general population, though a recent World Health Organization report stated:

> No conclusions could be reached on the average indoor ambient temperature below which the health of the general population may be considered endangered... For certain groups, such as the sick, the handicapped, the very old and the very young, a minimum air temperature of 20°C is recommended. (WHO 1987, p.19)

Work undertaken for the BRE reached the following conclusion on the adverse effects of temperature:

> It is not possible to define a 'safe limit'. The lower the temperature the greater the risk to the individual. Discomfort is perceived by most people when the ambient temperature is less than 18°C but adverse health effects have not been identified until the temperature declines to less than 16°C. Below this temperature there may be an increased risk of respiratory infection, one of the causes of increased winter mortality. (Mant and Muir Gray 1986, p.5)

It is not unreasonable to deduce from this statement that the start of discomfort is likely to indicate the commencement of health risks, so that the temperatures required for comfort and for health are broadly the same. That is the assumption made here: for comfort and health, the temperature of an occupied room should average 21°C.

One group who may need higher temperatures are the sick and disabled. Restricted mobility inevitably results in more time spent in the home and the reduced level of activity means that a higher temperature is needed to achieve comfort. Long-term sickness or disability in pensioners, or that results in the inability to work in younger people, are covered by specific social security benefits, but many people who are not officially recognized as coming into these categories report themselves as suffering. With illness, self-diagnosis is probably a good indicator of behaviour, for instance of activity — or inactivity — within the home. Long-term ill health affected people in nearly half of all SB claimant families in Berthoud's survey (1984, p.A3a). The problem is greatest for pensioners, two-thirds of whom reported a long-term illness, disability or infirmity which limited activity (ibid, p.A3). In non-elderly families, 38 per cent of poor families reported that the head of the household was temporarily or permanently out of work due to sickness or because of a limiting, long-standing illness, in comparison to 24 per cent in the general population (Berthoud et al. 1980, p.37). Thus, there is an association between poverty and ill health in all age groups and the link is strongest for pensioners.

There is another consideration with regard to elderly people. They have 'a reduced ability to sense the cold' and a variation of 4-5°C is barely noticed, whereas a 1°C range is recognized by younger people (Collins and Exton-Smith 1983, p.520). This means that policies to help the elderly keep warm should focus particularly on the provision of warmth, rather than income support: if they cannot recognize that they are cold, the old are unlikely to spend sufficient money on heating. From Chapter 3 it is known that the elderly are unlikely to incur fuel debts, though they often have disproportionately high fuel expenditure and are the group most likely to be using general tariff electricity for heating (Chapter 5). This combination of careful budgeting, energy inefficient heating and temperature insensitivity begins to explain why low-income pensioners have cold homes and a high morbidity and mortality risk in winter.

'In the UK it is usual for heating to be switched off at night' (Uglow

1981, p.6). This means that some households experience very low temperatures at night, particularly those where the rate of heat loss is high and the house cools down quickly. Because low temperatures can be harmful to health, 'some households heated their bedrooms overnight for the benefit of children' (Campbell 1985, p.35). As stated, when the temperature of the air falls below 16°C there is an increased risk of respiratory disorders, regardless of the level of insulation provided by bedding. If the risk of respiratory disease is limited to groups, such as the elderly, young, sick and disabled, then it is only these vulnerable people that need a minimum night-time temperature of 16°C in bedrooms. However, the risk may be more widespread so that a minimum temperature of 16°C at night needs to be maintained for health purposes in many occupied bedrooms.

Condensation control

The absence of adequate temperatures within the home can be demonstrated by damage to the fabric of the property. For instance, 'the snow and freezing temperatures in January and February 1985 ... cost over £150m' in extra household insurance claims (ABI 1986, p.31), and the cold spell in January 1987 resulted in frozen or burst pipes and heating systems in up to 50 per cent of the properties of some housing associations.

Another indicator of poor heating is the occurrence of condensation, as the moisture content of warm air condenses out on a colder surface. When the relative humidity is above 70 per cent, mould grows in damp areas and creates a health hazard for people, particularly children, with allergies or asthma (Burr 1986; Hunt *et al.* 1986). The number of occupants, level of heating, ventilation rate and insulation in the building fabric all contribute to causing and controlling condensation and demonstrate the complex interrelationship of factors involved in achieving a satisfactory home environment in winter.

> Severe dampness affects approximately 2m dwellings in the UK, with a further 2.5m affected to a lesser extent. The problem occurs mainly in rented housing, both public and private, and has been attributed largely to condensation. (Bravery *et al.* 1987, p.1)

In these households, there is mould growth, damage to decorations, floors, carpets or furniture (ibid, p.2). The largest number of the families with severe condensation live in public sector housing (0.9m), though the risk is greatest in the privately rented sector (29 per cent affected):

> We moved in the house in the summer and everything was dandy until the weather changed to winter. It was like living in a perpetual draught. Then the ceilings started to drip on us. The walls streamed with water, the ceilings were just one wet mass. There was mould (green and black) everywhere — in cupboards and wardrobes, carpets, curtains, furniture, all our clothes were ruined, absolutely everything. (Mrs Blood, quoted in NCC 1979, p.14)

In this and other cases, the courts are holding landlords responsible for repairs to combat damp, condensation and inadequate heating (as reported in Chapter 5). The necessary work usually involves both insulation and improvements to the heating system to produce lower running costs for the tenants. Increased ventilation, although the traditional advice of housing managers, is not an adequate solution as the lower level of relative humidity is accompanied by a colder dwelling. Similarly, dampness in the building fabric requires additional heat in order to evaporate it, so that condensation demonstrates heating problems and causes further ones.

The temperature required to prevent this depends on the moisture generated by the occupants, as Milbank states:

> For a given house the minimum heat requirement will change with the number of occupants. In general more occupants lead to more moisture generation and hence higher minimum heating requirements to control humidity levels and reduce mould growth. (1986b, pp.45-6)

There is little further guidance about the temperature required in rooms that are rarely used in order to avoid condensation, though a MIT of 14°C was necessary in a house with little ventilation and five occupants (Boyd *et al*. 1988). The only effective cure is the provision of adequate heating in insulated buildings — improved insulation alone did not prevent the formation of mould in unheated rooms (Milbank *et al*. 1985, p5). The objective is to have policies that enable people to keep healthy. As there is a risk of condensation and mould growth in unheated rooms, the provision of healthy housing seems to indicate that whole house heating is needed, regardless of the size of the family.

Recommended standards

Long before research was undertaken in climate chambers, opinions were being expressed about comfort temperatures:

> There is little conflict of opinion about the range of the comfort zone for winter warmth. For general purposes a fairly uniform air temperature of 65°F (18°C) has been accepted in this country for over a century ... as recommended by Reid in 1844. (Cmd 6762 1946, p.48)

The comfort zone defined by Collins (above) starts at 18°C and this temperature would provide comfortable conditions if people were wearing slightly thicker and non-synthetic clothes, as seems likely in 1844 and 1946 and had the same level of activity as now (e.g. 1 clo in Table 6.2). Therefore, whether or not we actually have warmer homes now than our predecessors did, there appears to have been no marked change in our desired temperatures, except as they are affected by our clothing levels.

Recommended design standards for domestic heating (Table 6.3) have used comfort temperatures (similar to those in Table 6.2) and assumed a relationship between activities and specific rooms, that may not always be valid. For instance, 13°C in the kitchen is based on a very active level of

Table 6.3 Temperatures recommended in design standards (°C)

	1961-81	1968-76	1977	1983	1983	1983
Living rooms	18	21	21	21	20	21
Kitchen	13	21	18	16	18	21
Bedrooms	18*	21	18	18	18	21
Bathroom		21	18		18	21
WC		21	18			21
Halls and passages	13	21	16	16	18	21

Sources and *notes*:   (1)   (2)   (3)   (4)   (5)   (6)

(1) * if use levels require. Recommended in Parker Morris Report (1961); made mandatory for public sector housing, except for bedroom heating, in MOHLG circular 36/67, effective from 1 Jan. 1969. Not required in local authority dwellings since 1980, though still the standard for housing associations.
(2) Standard for housing for the elderly in Design Bulletin (MOHLG 1968), MOHLG circular 82/69, DHSS guidance notes (1972). All withdrawn by November 1976.
(3) British Standard 5449 (1977) for forced circulation, hot water, central heating systems for domestic premises: to be achieved when the external temperature is −1°C.
(4) The Institute of Housing and RIBA (1983) — family housing.
(5) Energy Efficiency Office (1983).
(6) Institute of Housing and RIBA (1983) — minimum for elderly people.

There is a general vagueness as to when these standards should apply, beyond 'when the building is occupied' (BS5449). For instance, there is virtually no guidance on night-time temperatures.

food preparation; meals could not be eaten in this environment in comfort. Discomfort would also occur if a bedroom at 18°C was used for homework or other sedentary activities. These Parker Morris temperatures were those used in public sector housing and demonstrate that they would inhibit families from using all the space within the home, during the day. It is recognized that desired temperatures in the home are increasing: in the BREDEM model (Glossary), the three demand temperatures are 17°C, 21°C and 25°C corresponding to cool, average and warm houses (Anderson *et al.* 1985a, pp.21, 24). The figure of 21°C represents average comfort conditions (based on time-budget diaries) and is consistent with other standards and not too high.

The information provided so far can be used to define a heating standard — a benchmark against which to judge whether people have adequate warmth or not. The temperatures recommended in this heating standard are:

— 21°C in occupied rooms during the day for health and comfort; a slightly higher temperature is needed for those made inactive by sickness or disability;

— 14°C in unoccupied rooms to prevent condensation (day and
   night);
— 16°C at night in bedrooms occupied by vulnerable people, to
   prevent the risk of respiratory disease.

The resultant MIT can be established when the hours of heating have
been fully considered with other occupancy issues, below.

This heating standard clarifies an issue raised by the EEO:

> Consumers have diverse and non-economic objectives in discretionary
> expenditure — entertainment, comfort, leisure interests etc. — and are
> not necessarily motivated by the prospect of reducing energy costs.
> (HC 262 1986, p.xii)

How much warmth is necessary for health and at what point does
additional warmth constitute discretionary expenditure on comfort — a
peripheral part of modern standards of living? The evidence presented in
this chapter is that the difference is in the range of 2-5°C, depending on
clothing and activity (i.e. from 16°C up to 18/21°C). To some extent, the
above statement by the EEO represents an own goal. If the value of the
additional warmth is greater than the value of the energy savings, then
consumers are demonstrating that they consider adequate warmth is
important, rather than discretionary. Although people will not die or get
ill as soon as they start to live in cold conditions, the risk is increased. A
standard of heating that is below comfort temperatures would exacerbate
that risk.

External temperature — COWI 5

The amount of heating to be purchased to achieve a given temperature is
affected both by the temperature outside the house and by the amount of
heat gained inside the home from incidental sources, such as the sun,
people and non-heating energy use. There are several aspects of the
external climate that affect the cost of keeping warm: ventilation rates
increase with wind strength and rain-dampened building fabric has a
higher heat loss. Markus (1982) has argued that a Climatic Severity Index
is needed to assess the impacts (for instance on energy consumption) that
result from the combined effect of climatic factors, though this would
depend upon better published data. The influence of rain and wind are
not considered further in this chapter; the emphasis here is on the
external temperature and the moderating influence of sunshine.

There are two ways in which the external temperature can be measured
— the degree and the degree day (see Glossary). Degree days (DD) are
used in most official publications and data sources as they facilitate
external temperature comparisons between regions of the country and
over time (EEO 1984). DD are calculated to a base figure (usually 15.5°C in
Britain, but not necessarily in other countries), because it is assumed that
incidental gains within the house are equivalent to 2.8°C, to give a final

internal temperature of 18.3°C. A 24-hour MIT of 18.3°C is, therefore, the definition of comfortable implicit in any British DD calculation.

This fixed base for DD implies that in an average house (where the heat loss is 295 W/°C — Table 4.6), the incidental gains total 71 MJ in a 24-hour period. This is a high rate of gains for a domestic household, as shown below, probably because the original data were obtained from office consumption. Where incidental gains are lower, or heat loss rate higher, DD to a fixed base of 15.5°C underestimate the need for heating; the converse is equally true. For accuracy, the base for DD has to be calculated in relation to the heat loss and incidental gains of the building, particularly in low-income homes.

One advantage of DD as a measure is that stating the length of the heating season becomes unnecessary. By definition, the occurrence of a DD indicates that incidental gains are insufficient to keep the building warm, and additional heating is needed. Defining the length of the heating season otherwise is a fairly arbitrary process, though the nine-month period, September-May inclusive, is usually taken now (EEO 1983, p.12).

Degree days provide a quantitative basis for comparing different regions of the country. For instance, in an average year, the north-east of Scotland with 2908 DD has a climate that is 47 per cent more severe than the south-west of England with 1972 DD (Figure 6.1). However the variations between years are even greater: over the last 20 years, the coldest December had 89 per cent more DD than the warmest December, averaged across the whole country; for January the range was 73 per cent (based on data from the Meteorological Office). Because of these annual variations, the DSS has devised the exceptionally severe weather (ESW) policy to assist claimants. As the COWI formula indicates, a 10 per cent increase in the number of DD (and therefore decline in external temperature) requires a corresponding (but smaller) increase in fuel consumption costs, if all other factors are kept constant. Furthermore, where the energy efficiency of the house is low, and the cost of keeping warm high, this additional cost is greater — 10 per cent of a large amount is more than 10 per cent of a small amount. The cost impact of cold weather is inevitably greatest in energy inefficient homes.

Other regional variations are difficult to establish, though the proportion of households using gas has a clear spatial distribution (Figure 6.1). The figures are taken from the FES and do show an annual variation. Scotland recorded 62 per cent of households using gas in 1985; the 9 per cent drop in 1986 probably indicates sample bias. However, about 45 per cent of the homes in the coldest region have less energy efficient homes, because they are not using one of the cheapest fuels. Northern Ireland is slightly warmer, but with no natural gas (only town and calor gas) has extremely high fuel bills. It is not possible to extend this analysis, but it does indicate a greater vulnerability to fuel poverty in some areas, like Scotland, either because of the severity of the climate or because of the lack of the least expensive fuels, or both. There are no regional variations in the amount of money paid in benefits to offset these disadvantages.

Figure 6.1 Degree days and gas ownership, UK 1986

Note: Percentages are the households in the region (not gas area) that record gas expenditure in the FES. Numbers are degree days.

Sources: EEO 1984, p.16; Hutton and Hardman 1990, Table 8.2

Incidental gains

There are three main contributors to incidental gains: the energy used for non-heating purposes releases heat as a by-product and thus contributes to raising the temperatures within the house. Some temperature gain also comes from the presence of people and from the effect of the sun shining into the house, and these second and third categories are 'free'. There are two points to be made about non-heating energy use. First, because it has to be paid for, the level of consumption and, therefore, of incidental gains, in a low-income house may be much lower than in a better-off household. The range of consumption for the non-heating uses is not well understood, but is probably as considerable as some of the other factors influencing the cost of keeping warm. For instance, Nicholls and Rees found that hot water for baths varied from 6 litres a day for a 'frugal pensioner' to 140 litres, per person, per day where the solid fuel CH produced 'excessive amounts of hot water' (1983, p.8). Secondly, the large proportion of the energy emitted as heat (eg lighting and cooking) is only contributing to useful gains during the heating season; during the summer the heat is unwanted and wasted and an indicator of the low efficiency in the intended application. Only 3 per cent of energy is converted into light by an ordinary bulb (Evans and Herring 1989, p.9).

Gains from solar radiation are one of the sources of incidental gains, though of relatively minor importance. For instance, Penz has estimated that in an average house (not well insulated), heated to a high standard, but randomly oriented, the average solar contribution is only 10MJ a day, throughout the heating season — about 6.7 per cent of the useful heating load as he calculated it (1983, p.94). More recently, Milbank has stated that 16 per cent of the space heating requirement in the existing UK housing stock comes from solar energy (1986a, Figure 1). The variation between these figures may be because Penz's is a proportion of a high standard, whereas Milbank is using present average consumption. The contribution made by solar gains depends upon the family's occupancy pattern and demand temperatures. In a home that is well heated only in the evening, solar gains are of minimal contribution, whereas they are a substantial source of warmth to the occupants of a south-facing room on sunny days. Unfortunately, solar gains cannot be guaranteed or relied upon.

All the warmth created by personal activity in the home naturally reduces the need for heating. For a person in the home during the day (13 hours — Table 6.2), the amount of energy expended is equivalent to 5MJ. This raises the temperature by about 0.3°C, and is of negligible importance unless there are several people in the same room, or in the house. Metabolic output is easy to quantify if the occupancy rate is known for each person in the family, with average activity patterns, using the data on metabolic rates in Table 6.2.

In practice, at the moment, the incidental gains occurring in low-income families are unlikely to be as high as those assumed in the DD formula. The total benefit from incidental gains is calculated by Campbell to be 40-50 per cent of the total winter heat requirement in modern houses occupied by large families (1985, p.13), partly because the temperatures were quite modest. However, in an older property, occupied by a single person heating the living room only, the incidental gains from body heat, lighting and TV could be as low as 7MJ a day, barely 5% of the heating requirement.

At low temperatures, the proportion of gain from incidental sources is large — in the extreme, when the heating system is not used, any temperature gain indoors comes entirely from non-heating sources. Similarly, at low rates of heat loss, the incidental gains are better utilized (lost more slowly), and form a high proportion of the total heating load. Thus, it is the absolute level of benefit that is of importance — a high proportion can mean either low temperatures or low heat loss, and occurs in two quite different situations. If incidental gains are minimal, as in many low-income households, the DD base used in energy audits should be raised to, say, 17°C instead of 15.5°C in compensation, increasing the average number of DD from 2,500 to 3,000 (Anderson *et al*. 1985a, p.49). The lack of information on incidental gains in low-income households inhibits the development of accurate computer energy audits for these families.

Occupancy patterns — COWI 7

Rooms need to be warmest when they are occupied, so that the time that people are in the house is an important factor in a heating standard. Similarly, the activity within the house determines the number of rooms to be heated. This section examines the evidence on temporal and spatial occupancy of the house to define these parts of a heating standard.

Hours of heating

The information from the time-budget survey on the hours of occupancy has been given (Table 6.2). As the diaries were kept by individuals, it is not possible to determine the combined heating needs of the whole family, so the data provide only a minimum heating demand for the household. Hunt and Gidman established in their national survey in 1978 that 74 per cent of homes are normally occupied during a weekday (1982, p.108). The proportion in low-income homes is likely to be higher as a result of increased unemployment and the concentration of elderly households amongst the poor. On average, household income, size of family and number of workers increase together (see Table 6.4 below): only 2 per cent of households in the lowest income quintile contained someone who had been employed for the whole year (*Social Trends* 1985, p.88). Therefore, the typical low-income house is occupied throughout most of the day, requiring heating for 13 hours during the day and 8 hours at night: it is empty for a maximum of 3 hours. The activity patterns of the unemployed, retired, housewives (with or without young children), sick and disabled are remarkably consistent in this respect. The activity pattern for people at home does not vary much according to the day of the week, although it does for workers.

A further analysis was undertaken by tenure and this extended the range in hours of occupancy: someone in full-time employment, renting from the council, spends 6 hours a day awake and in the home, whereas, in the same tenure, a housewife, without a young child, spends 14.6 hours at home in need of heating (Figure 6.2) — a multiple of 2.4. Not all the people who spend most of the day at home are poor, but an 'adult at home' is a characteristic of 62 per cent of the households with large fuel bills (Table 3.1).

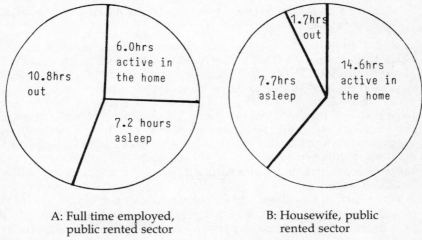

A: Full time employed,   B: Housewife, public
   public rented sector   rented sector

Figure 6.2 Maximum and minimum hours in the home, by group

Source: Boardman 1985

For people to be warm, each space needs to be warm when it is entered — rather than achieving that temperature some time later — which means pre-heating rooms. Similarly, it may be possible to turn off the heating before leaving the room, so that the hours of heating and occupancy time may be the same, though not concurrent. The exact relationship between the hours of occupancy and the hours of heating depends on the type and responsiveness of the heating system. A more detailed pattern for comfort temperatures cannot be established, because the blocks of time spent on different activities is not known. For instance, most non-workers — except housewives with young children — spend at least 7 hours a day at seated tasks. These could be spread throughout the day, or occur mainly in the period 2-10pm — Hunt and Gidman found a consistent decline in mean activity rate throughout the day, particularly for women (1982, p120). Thus, a temperature gradient, rising throughout the day, might provide the greatest comfort, whilst still achieving the 21°C daytime average.

Spatial heating and family size

How many rooms are to be adequately warm? A policy aim of one warm room is effectively suggested in the new fitness standard and some Electricity Boards are promoting the same concept through Budget Warmth (see Glossary). Yet in many homes with CH all the rooms are warm. In neither case is the level of heating apparently related to the size of the family. The implication of a one warm room policy is that the health of the occupants is at risk whenever they use other parts of the dwelling:

> I can't afford to keep the house warm. There was ice inside the flat last winter. I shut up the whole flat and just live in the bedroom in winter. (Elderly private tenant, in Smith 1986, p.35)

A way of defining the relationship between the amount of space to be heated and the numbers in the family is to use the basic tenet of the Parker Morris Report (MOHLG 1961) that privacy is required during the day, as well as at night, and this can only be achieved if there is both adequate space and adequate heating. The amount of space needed to avoid overcrowding has been defined since the Housing Act 1957. Known as the bedroom standard (see Glossary) it is used by local authorities to match tenants and housing, by giving, for instance, separate bedrooms to children of different sexes over 10 years (Smith 1977, p.66). Thus, care is taken to provide families with sufficient space, for a variety of social and health reasons.

The principle has already been suggested that the temperature in an unoccupied room is lower than that required for comfort and health, but is sufficient to prevent condensation. But, with small families in large dwellings it is not easy to define what is an unoccupied room. For instance, it could be suggested that for someone to move about performing normal household activities in comfort during the day the living room, bedroom, kitchen and bathroom need to be kept warm, with an additional room heated to comfort standards for the second and subsequent occupants. Thus, a family of two adults and two teenagers would be able to have five rooms (plus kitchen and bathroom) at 21°C when they are all at home.

The effect of this heating standard is shown in Table 6.4. If it is assumed that non-worker members of the family are in for 13 hours a day, heating the minimum space, as just defined, to 21°C and workers are in for 7 hours heating a further number of rooms, the total number of room hours of warmth (at 21°C), per income group can be derived. Because increasing income is correlated with more members in the family, the richest 70 per cent of families have the largest heating need, though this is only 11 per cent more than that of poor households. In comparison, the richest 70 per cent spent 34 per cent more on energy than the poor in 1988 (£11.34 : £8.48 from Table 3.4). Thus, on average, poor households contain less people

Table 6.4 Heating demand per income group, UK 1988 (warm-room* hours/day)

Income decile	Persons per household			Heating demand		
	Total	Non-workers't	Workers	Non-workers	Workers	Total
1	1.179	1.105	.074	53.37	.52	53.9
2	1.824	1.657	.167	60.54	1.17	61.7
3	2.133	1.762	.371	61.91	2.60	64.5
4	2.176	1.472	.704	58.14	4.93	63.1
5	2.530	1.376	1.154	56.89	8.08	65.0
6	2.850	1.460	1.390	57.98	9.73	67.7
7	2.970	1.282	1.688	55.67	11.81	67.5
8	3.021	1.179	1.842	54.33	12.89	67.2
9	3.140	1.071	2.069	52.92	14.48	67.4
10	3.340	1.120	2.220	53.56	15.54	69.1
All	2.516	1.348	1.168	56.52	8.18	64.7
1 elderly	1	1	0	52.00	0	52.0
2 elderly	2	2	0	65.00	0	65.0

Notes:
* Two rooms, kitchen and bathroom are heated for the first non-worker and an additional room for each subsequent non-worker for 13 hours a day. Workers are assumed to be in the house for 7 hours. All occupied rooms heated to an average of 21°C.
† Non-workers include children, whether at home or at school.

Source: Persons per household — FES 1988, p.6

and so require marginally less warm space than richer households (in absolute numbers of room-hours). Large, poor families have the greatest need for heat as they have high spatial and temporal demands. As the poor are in the home more during the day, a larger percentage of their home needs to be at 21°C, so that the MIT in the homes of low-income households should be higher than those of other families.

The data in Table 6.4 shows a theoretical relationship — in practice the size of families and their dwellings do not fit so precisely. And you cannot have 0.1 of a room, any more than you can of a person. In addition, because of the need to keep the remaining rooms free from condensation, even where few people occupy a large dwelling, all the rooms are provided with at least background heating. In effect, therefore, whole house heating is being recommended with two temperature zones: a higher temperature in occupied rooms, usually the majority or whole of the house, and a lower level in unoccupied or rarely used rooms.

There are technical and economic reasons why whole house heating provides the best approach and value for expenditure. With some heating systems, reducing the space to be heated to some portion of the whole would prove extremely difficult. This may be a criticism of the way engineers design systems and the need to provide for more sensitive spatial control, but for users of existing systems, such as underfloor electric heating, there is no choice. Further, the interrelationship between the capability of the heating system, heat loss within the home and variations in desired temperatures is quite complex. Reducing the space heated does not result in comparable energy or money savings (Figure 6.3). One reason for this is that the internal partitions in the dwelling are not designed to retain heat efficiently, so that there is always a high rate of heat loss between heated and unheated parts of the same home. A similar principle applies with heating systems: the fuel economies that result from intermittent use are not as great as the reduction in heating time (Markus and Morris 1980, p.354). Thus, to obtain a small saving in expenditure on heating requires a much greater proportional reduction in the extent and/or duration of the heating; only marginal savings are achieved by partial heating. ETSU have calculated that a 45 per cent reduction in heating hours results in a 15 per cent reduction in energy consumed (based on Evans and Herring 1989, p.55). The 11 per cent variation in warm-room hours between the poor and other families (Table 6.4) would, therefore, result in low-income households consuming 4 per cent less energy, if other factors were equal. Whole house heating, therefore, represents the spatial distribution of warmth included in the target heating standard.

As Figure 6.3 demonstrates, the crucial variable is the level of insulation: partial temporal or spatial heating in a well insulated house can produce higher temperatures than extensive heating in a poorly insulated one. It is the complexity and sometimes non-linearity of the interactions between space, time, temperature, cost and heat loss that make simplified models, like COWI, less accurate than computer-based audits.

When these temporal and spatial standards are combined, the proportion of the dwelling that is heated to 21°C varies according to the number of people in the home. A heating standard that is related to the number of occupants accommodates the point made by Milbank earlier, that more people in the home result in the generation of additional moisture and the need for extra warmth to prevent condensation. Table 6.5 provides some information to explore the implications of this recommended heating standard further. The bedroom standard, mentioned above, indicates which households have large dwellings for the number of people in the family. Households living at densities above the bedroom standard are, by this definition, under-occupying their dwelling. As can be seen from Table 6.5, there is a considerable range across the tenures, with housing association tenants most fully occupying their properties and outright owner occupiers the least. A closer data match between income, dwelling size and number of people cannot be established. The greatest problems are experienced by the 20 per cent of people in privately-rented unfurnished accommodation who are trying to

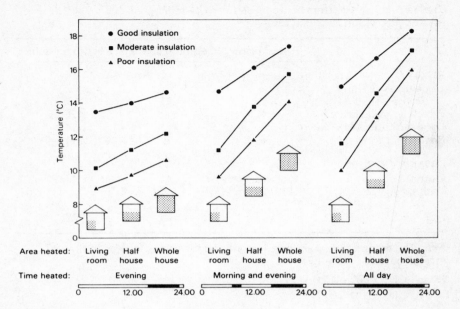

Figure 6.3 Average house temperatures under different heating regimes

Notes: The shaded part of the house and the black sections on the bar indicate the space heated to 21°C and the time period. The three insulation levels range from none (solid walls, 2 ach) to good (1982 Building Regulations, plus double glazing and draughtstripping). The temperature plotted is the resultant 24-hour whole house MIT.

Source: Anderson *et al.* 1985b, p.2, BRE Crown Copyright 1990

heat fairly large dwellings on some of the lowest incomes (Table 4.7). With the exception of this sector, largely oversized dwellings (>1 room above) do not appear to be a problem for many poor households. Thus, recommending whole house heating in the heating standard would not imply substantial numbers of warm, underoccupied dwellings and a misuse of resources, particularly because of the marginal increase in costs that results from raising the temperature in one or two rooms from 14°C to 21°C.

The amount of per capita space is greatest in small households (Table 6.6). For instance, there were 1.4m single person households in 1981 living in five or more rooms, excluding the kitchen and bathroom (1981 Census, Table 29), though their income levels and age-groups are not known. It is suspected that many of these households are elderly owner occupiers, some of whom are on a low income. One of the reasons for fuel poverty may be low-income pensioners living in large properties, that they are unable to heat or maintain properly. The solution to this situation involves both sensitive assistance to those wishing to stay and a wider

Table 6.5 Difference from bedroom standard by tenure, GB 1987 (% of households within each tenure)

	At or below	1 bedroom above	>1 bedroom above
RENTED			
Housing association/ co-operative	72	21	7
Privately, unfurnished	41	38	20
Local authority/New Town	50	36	14
Privately, furnished	69	20	11
OWNER OCCUPIED			
Outright	15	41	43
With mortgage	29	44	27
ALL	33	40	27

Note: The tenure groups are listed in order of increasing income for lower income quartile (Table 4.7)

Source: GHS 1987, p.123

Table 6.6 Relationship between number of people and number of rooms per household

	England and Wales 1981		GB 1988
People	Rooms	Average occupancy (persons/room)	% of households
1	3.95	0.25	26
2	4.79	0.42	34
3	5.23	0.57	16
4	5.61	0.71	16
5	5.80	0.86	5
6	5.91	1.02	
7	5.94	1.12	2
8+	6.14	1.30	
	4.95	0.54	
Average persons/household	2.7		2.5

Note: A room excludes bathrooms, WCs and small kitchens. 1981 and 1988 data cannot be combined

Source: Col 1 and 2: *Social Trends* 1985, p.124; Col 3: *Social Trends* 1990, p.36

choice of alternatives for those who would like to move (Chapter 10). Because the number of people in the average household is declining, whereas there is little new building, there is an inevitable growth in the amount of space occupied by each person. For instance, in 1981 22 per cent of all households contained only one person; by 1988 it was 26 per cent.

Very little is known about the proportion of the house that is heated in practice, and certainly not about the relationship between that space and the number of occupants. In the National Fuel and Heating Survey of 1,600 households in 1975, the poorer families heated 2.9 (53 per cent) of their 5.5 rooms, whereas other households heated 3.8 (62 per cent) of their 6.1 rooms at least occasionally (Field and Hedges 1977, p.71). More recently, the 1981 English House Condition Survey found that 31 per cent of households regularly heat less than half their rooms, and a further 16 per cent about half (DOE 1983, p.68). In the latter case, the definition of a room excluded bathrooms and small kitchens. The energy data from the 1986 English House Condition Survey should provide more information when it is published (late 1990), though the poorest families undoubtedly continue to heat the least space, if only because of the inadequacy of their heating systems.

Conclusions

A heating standard is proposed to identify the amount of warmth required in a home if the occupants are to be comfortable and healthy and not to suffer from condensation. Low-income households are known to be in the house most of the day, rather than out at work, so the proposed standard is:

a temperature of 21°C in two rooms, kitchen and bathroom for 13 hours during the day for the first occupant, and an additional room for each person when they are in the home. At night, 16°C in the bedrooms of people vulnerable to respiratory disease. Other unoccupied rooms to be maintained at 14°C to prevent condensation, mould growth and, thus, ill health. This gives a 24-hour, whole house MIT of about 18°C.

This is, in effect, whole house heating, with different temperatures zones — the basis of most computer programs, such as those based on BREDEM (see Glossary). Energy audits use the DD formula which is calculated to achieve a MIT of 18.3°C in Britain — similar to the proposed heating standard. However, the DD formula assumes a greater level of incidental gains than is appropriate for most low-income families.

The three COWI factors introduced in this chapter show much less variation than the previous four. Strong personal preferences in comfort levels are rare — when free of economic constraints, 95 per cent of people chose a temperature within a narrow band. Similar activity levels when in the home mean that the optimum demand temperature can be taken as a constant 21°C, whilst in the home. The poor spend most of their day in the home, effectively doubling the hours of heating they need over that of someone out at work. The third factor, external temperature, shows a regional and larger annual range. Cumulatively, these three COWI factors quantify the amount of warmth needed by a household and prove that it is the cost of the warmth that shows the real variation.

Warmth is a commodity that is highly valued: heating the living areas was classed as a necessity by 97 per cent of households in Mack and Lansley's survey of minimal living standards, so that it achieved first place in the rankings (1985, p.54). Therefore, temperature variations occur because of differences in the cost of warmth or economic constraint, not because people prefer to be cold. This simplifies an assessment of fuel poverty by reducing the main variables to energy efficiency and income.

To avoid ill health, certain groups in the population need their warmth more than the rest. For instance, elderly people become less able to judge cold, even though it may be harming them. In addition, the elderly (and other low-income families) may have low expectations:

> There may be some, especially among the elderly, who say they have heating in their living areas, when in fact they can afford to have it on only for an hour or so in the evening, and usually wrap up in rugs or go to bed to keep warm. Because of low expectations, they do not feel that they are forced to go without heating, but this does not necessarily mean that they are not deprived. (Mack and Lansley 1985, p.105)

This demonstrates clearly the difficulties of depending upon the elderly, in particular, for an assessment of their needs and, therefore, of appropriate, self-generated action. Schemes such as Budget Warmth (Glossary) and the Glasgow heat-with-rent (Markus 1987, p7) which involve fixed prepayments for heating are good solutions, provided they are not too expensive. But the real implication for policy is that those vulnerable to the cold, such as the sick, disabled, elderly and young need to be able to purchase inexpensive warmth, through energy efficiency improvements.

7 Achieved warmth

As already described, there is scant recognition in building and environmental health standards in Britain of the relationship between adequate warmth in the home and good health. The lack of official interest in warm homes is reflected in the lack of data on temperatures in occupied homes, especially on the extent of cold living conditions. There have been only two national surveys, of different populations (Wicks 1978; Hunt and Gidman 1982). The second of these is referred to extensively here, because, despite the gap of more than ten years since the survey took place, there is no more recent data. Temperature measurements in poor homes formed part of the 1986 English House Condition Survey, though these are not published yet.

Data on temperatures in the home can be obtained from measured surveys or deduced from useful energy consumption (Chapter 8) to provide information on the conditions in low-income homes. The evidence in Chapters 4 and 5 about energy efficiency levels indicates wide variations, to the detriment of the poor. These findings are confirmed if it can be shown that low-income households are suffering from the effects of energy inefficiency by both being colder and having to purchase more expensive warmth than other families.

Because of the shortage of temperature data, the impact of cold housing conditions has to be measured partly in terms of excess winter deaths (EWD). These are set in context through comparisons with other countries using data on temperatures inside and outside the house, as well as building standards. These enable the role of different aspects of the external climate to be examined. First, some data are provided on the temperatures recorded in the homes of households that are not poor, to provide guidance about current expectations and to validate the suggested heating standard.

Interpreting temperature surveys

This section could have been headed 'When is cold cold?', because it addresses the relationship between partial heating and the MIT. When a dwelling is partially occupied and heated, to comfort conditions, the MIT could still be low, because of the influence of cooler, unoccupied rooms (Figure 6.3). Thus, low temperatures and comfort conditions are not mutually exclusive. Similarly, a high MIT in a rarely occupied home represents a waste of energy.

The highest 24-hour MIT should be found in an energy efficient home, occupied during the day, and the lowest MIT in an underoccupied dwelling which loses heat rapidly. A low level of occupation occurs when someone is in the house for few hours a day, or is living in only some of the rooms. Thus, there is a fairly complex interrelationship between occupancy patterns and average temperatures. In the previous chapter it is shown that the homes of low-income non-workers should on average be the warmest (with the highest 24-hour MIT), because they are the most constantly occupied, even though they need fewer room-hours of warmth in total. In a fully occupied house, the family is warm if the 24-hour MIT is at least:

$$[(21°C \times 13 \text{ hours}) + (14°C \times 11 \text{ hours})] \div 24 = 18°C$$

In colder occupied homes, the space is not fully used, or there is discomfort, or both. It is much easier to determine when the household is warm than it is to be certain about the implications of lower temperatures. There is the further complication of whether all warmth is beneficial, particularly in relation to heat emissions from night storage systems. Another instance is when a house is well insulated and the temperature at night barely drops below the daytime level. These 'involuntary' temperature gains are determined by the physical characteristics of the house and may be of little extra value to the occupant (Pezzey 1984, p.183). It is important to differentiate between these two types of additional warmth: the unwanted output from a night storage heater has to be paid for and often replaced by an expensive alternative later in the day. The benefit of additional insulation is that it makes purchased warmth more durable and of better value — it is retained longer by the building fabric. Present data on measured temperatures is not sophisticated enough to distinguish between these two situations, so in this chapter and in Chapter 9, it has been assumed that all warmth is beneficial.

Temperature trends and averages

The first priority is to establish what is known about temperatures in average households, to compare these with the heating standard and then to examine what is happening by income group. As the majority of households have CH (Table 5.4), the temperatures in CH homes can be taken as those 'commonly approved by society' (Townsend 1979, p.31) and thus the reasonable aspiration of the poor. Data are available from small-scale research, designed to test developments in heating systems or insulation levels. Results from 1946-78, collated by Hunt and Steele (1980), show temperatures in CH homes rising, with the evening living room temperature in CH homes levelling off at just under 21°C in the early 1970s. Subsequently, it appears that living room temperatures are still rising and 21°C is considered an average and 25°C warm (Anderson *et al*. 1985a, p.21), over the whole day. According to Pimbert and Fishman:

to satisfy 90 per cent of the population a heating system should be capable of achieving 23°C in the late evening in the main living room. (1982, p.7)

Both of these confirm that the daytime average of 21°C in the main living rooms, as in the heating standard, reflects current practice rather than a maximum.

In practice, most houses, even CH ones, are considerably colder. In 1978, Hunt and Gidman (1982) took spot readings in every room of 900 homes, throughout Great Britain, in February and March. Visits were phased during the day, so no interviews or measurements were taken at night or when the house was empty. Thus, the average temperatures in the survey represent a 14-hour mean (9am-11pm), that cannot accurately be compared with a 24-hour MIT, but is close to the daytime part of the heating standard for families at home (21°C). In their survey, Hunt and Gidman found a 17.5°C 14-hour MIT in CH homes (ibid, p.112). Historically, temperatures in CH homes are known to have been rising at least until 1978 (Hunt and Steele 1980). There is no more recent measured evidence to confirm, refute or quantify the temperature trends in CH homes. NCH homes were known in 1955 to have a 24-hour MIT of 14.5°C (ibid, p.8), higher than the 14-hour, daytime MIT of about 14.3°C measured in 1978 (Hunt and Gidman 1982, p.112)

Using the BREHOMES model of the housing stock and national energy consumption, Henderson and Shorrock have *calculated* figures for internal temperatures. These are even lower than those measured in occupied houses (Table 7.1). The greatest confidence can be placed in the overall increase in average temperatures. The equal allocation of this increase between CH and NCH homes (always 2.5°C apart) is more questionable, particularly because of the lack of measured temperatures in NCH to provide any anchor points. As the authors correctly state, there is considerable scope for 'internal temperatures to rise before any practical upper limit is reached' (1989, p23). About 1.1°C of the average increase has resulted from households changing to CH. The rate of future temperature rises, therefore, depends partly on a continuing growth of CH ownership.

Table 7.1 Calculated temperatures in UK homes, 1970-86 (°C)*

	Centrally heated	Non-centrally heated	Average
1970	14.58	12.08	12.93
1986	16.46	13.96	15.76

Note:
* 24-hour average internal temperature during the six winter months

Source: Henderson and Shorrock 1989, p.23

The cold that is implied by a temperature below 14°C MIT can be gauged as it is below the figure at which respiratory disease begins (16°C), and undoubtedly includes periods of time when there is a risk of cardiovascular strain (below 12°C) — both figures from Collins (1986), quoted in Chapter 6. As it is not certain that temperatures are rising in NCH homes, the gap between the whole house 24-hour MITs for CH and NCH homes may be widening. Thus, the average British home is warmer, both because a growing proportion of dwellings have CH and because temperatures in CH homes are rising. The temperatures in

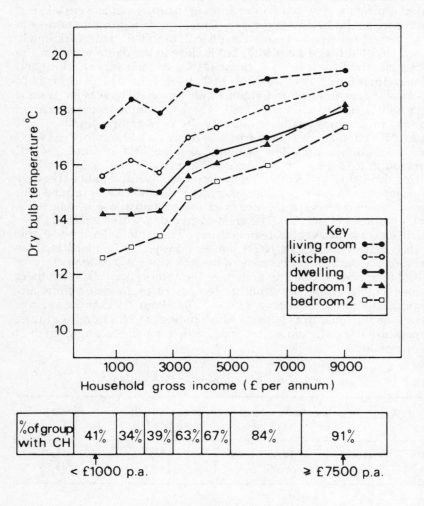

Figure 7.1 Temperatures inside dwellings by income, UK 1978

Source: Hunt and Gidman 1982, p.121, BRE Crown Copyright 1990

the remaining homes are still low and, perhaps, in relation to the average, colder.

These figures provide initial evidence of fuel poverty existing and worsening. They would have to be qualified if all NCH homes are under occupied, either because there are few people in large houses, or because NCH homes belong to workers who are out most of the day. The ownership of CH is strongly correlated with income (Figure 7.1), and so is the number of workers per household (Table 6.4). Therefore, low temperatures in NCH homes are not generally caused by empty houses during working hours.

Hunt and Gidman give no information on temperatures and size of family, but they do provide temperatures in relation to the number of rooms in the dwelling, with kitchens and bathrooms included as rooms. The coldest homes have five rooms, with both larger and, particularly, smaller dwellings being warmer (1982, p.114). They suggest no explanation, but for small dwellings, the greater warmth may indicate a higher level of occupancy, heating and use. With the larger properties, the result may demonstrate the ability to purchase more warmth, or high occupancy levels, or both. The proportion of households with CH increases linearly with the number of rooms, so that NCH homes are, on average, smaller.

As NCH homes are smaller, and more likely to belong to low-income families, there is no reason to qualify the statement that fuel poverty exists and is worsening, and is certainly found in NCH low-income homes.

In detail, the three lowest income bands (an unknown percentage of households) have nearly identical whole dwelling temperatures, and similarly low levels of CH ownership (Figure 7.1). In all income groups, the living room is the warmest room. However, in low-income homes, the between-room range of 12.6-17.4°C is more than double the range in high-income households of 17.4-19.4°C (ibid, p.122). Thus, the homes of the poor are 3°C colder on average and their warmest room had the same temperature as the coldest room in the home of a high-income household. Instead of being warm, the poor spend most of the day in cool or even cold rooms, whereas more affluent families are warmer whenever they are at home, and wherever they are in the house.

The level of heating in low-income homes, even those with CH, is still extremely low. In a survey of 50 low-income owner occupiers in London, the average living room temperature was 13.2°C and 9.6°C in the main bedroom, measured continuously over three weeks in the winter of 1984-5 (Meyel 1987, pp.26-7). As stated in Chapter 3, Mack and Lansley (1985) found that people ranked heating the living areas first in the necessities that should be attained for a minimum standard of living. This suggests that heating has a very high spending priority, so if the main rooms are cold this is a strong indicator of acute general poverty.

There are few other data sets which combine temperature measurements in ordinary homes with information about incomes. For instance, government-sponsored research in low-income homes is

usually to measure the effectiveness of insulation or other improvements
(see below and Chapter 9) and thus not informative about general living
conditions. A thorough, three-year investigation into many aspects of
condensation dampness in Glasgow included temperature surveys in
nine local authority houses, seven of which had household incomes
below £6,000 a year in 1983 (Markus and Nelson 1985, p.31). The average
24-hour temperature in winter for each family was in the range (ibid,
p.41-2):

bedrooms	5.4 - 16.4°C
living rooms	13.6 - 16.8°C
kitchens	9.9 - 17.3°C

(The coldest bedroom had a window that would not shut.) As Markus has
subsequently stated: 'Many readings of 8-14°C were recorded' (1987, p.6),
indicating cold homes and none that achieved the target heating
standard.

Effect of energy efficiency — building fabric

The remaining analysis of temperatures is in relationship to the main
factors identified in earlier chapters, starting with the energy efficiency of
the house and heating system. Internal temperatures are significantly
correlated with the age of the property (Figure 7.2): homes built since 1970
are 3°C warmer than those built before 1914 (the probability of this
occurring by chance is <.001 — Hunt and Gidman 1982, p.113). Part of
the reason for the temperature increase may be the presence of CH in a
higher percentage of the more modern homes. Similarly, the most
modern homes are occupied by households with higher than average
incomes (GHS 1987, p.126), but otherwise there is little correlation
between income levels and age of dwelling occupied. The dwelling
temperatures in 1914-39 homes are higher than in pre-1914 ones, despite
similar proportions with CH. The explanation appears to be that the
1914-39 houses have slightly better heat retaining characteristics, because
a higher proportion has cavity walls or added insulation, so that, as
shown in Chapter 4, the age of the dwelling can be used as a rough proxy
for the rate of heat loss. More modern homes are warmer partly because
they are more energy efficient.

A further example of the effect of increased energy efficiency occurs
when insulation is added. Using matched groups of houses, the
reduction in total heat loss can be plotted against the temperature rise
(Figure 7.3). All the work in these schemes was undertaken in public
sector housing and some other energy efficiency improvements to the
heating systems were also incorporated (the economic implications are
examined in Chapter 9). However, there is a clear trend towards higher
temperatures in better insulated dwellings — on average a reduction of 40
W/°C in the total heat loss resulted in a 1°C rise — indicating that families
desire greater warmth and will purchase it when the energy efficiency

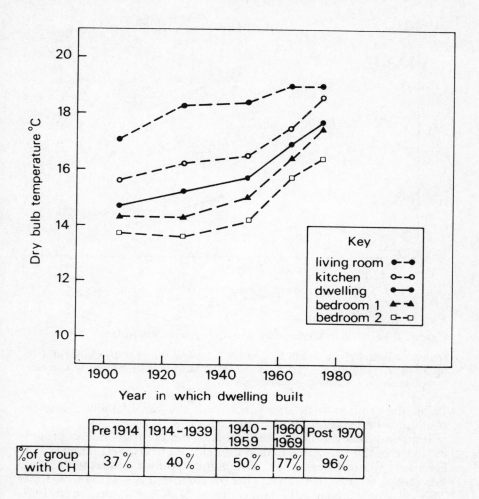

Figure 7.2 Variation of temperature with dwelling age, UK 1978

Source: Hunt and Gidman 1982, p.114, BRE Crown Copyright 1990

is improved. The houses had fairly low levels of total heat loss before improvement and even the warmest houses showed a gain in temperature.

The temperature in the dwelling is related to the original level of energy inefficiency and the amount that this is affected by improvements. In two government-sponsored projects in low-income homes, where the work undertaken was either draughtproofing and/or loft insulation (Hutton *et al.* 1985) or draughtproofing and some replacement of heaters (DOE

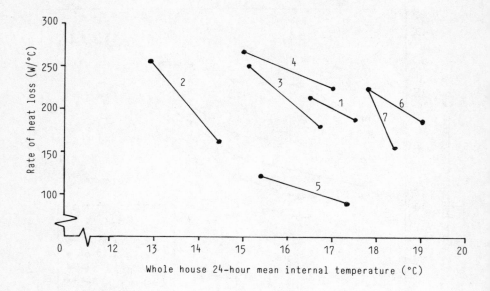

Figure 7.3 Variation in temperature with the addition of insulation, UK

Sources: 1, Birmingham — EIK Project 1982, Table 3.5a and b; 2-7 Campbell 1985, pp.7, 10; 2, Whitburn; 3, Hamilton; 4, Plymouth; 5, Darlington; 6, Coventry (single glazed); 7, Coventry (double glazed)

1978), the temperatures afterwards were still considerably below the conditions necessary for health and comfort.

Temperature levels in well insulated, owner occupied homes provide evidence of the likely demand in the homes of better-off families. The Electricity Council Research Centre monitored four houses with a heat loss rate of about 150 W/°C (half the British average) and found that the 24-hour whole house MIT ranged from 17.8-20.2°C (Brundrett 1987, Figure 4). The 24-hour MIT in the living rooms ranged from 19.5-23°C (ibid, Figure 3). Four estates at Milton Keynes with below average rates of heat loss (down to 135 W/°C), lived in by owner occupiers, recorded means of whole house temperatures, across the heating season, in the range of 17.8°C-19.1°C (Lowe *et al.* 1985, p.S18). The temperatures from both these reports are above that cited in the heating standard and indicate that the demand for higher temperatures in British homes has not been saturated and that the heating standard represents an average (or less than average), not a maximum level of heat. The uncertainty about higher temperatures is the extent to which they are dependent upon energy efficiency improvements and the extent to which they will occur, anyway, if real incomes rise.

Effect of energy efficiency — heating system

There has been only one study of measured temperatures in relation to a variety of heating systems — and that was undertaken in 1956-9 in local authority flats. The lowest temperatures were recorded where electric fires were used (14°C whole flat, mean weekly temperature) and the highest where the estates were centrally heated (i.e. district heating) with mean weekly temperatures up to 20°C recorded (Loader and Milroy 1961, pp.807-14). The siting of appliances affects the temperatures achieved. On one estate in Nottingham, the upstairs heating is provided by a night storage heater on the landing. At night, the bedrooms are cold and the temperature drops to 15°C because the doors are closed, whereas the temperature on the landing reaches 25°C, demonstrating 'a fundamental lack of understanding about how people live' (Nottingham Heating Project 1985a, p.6).

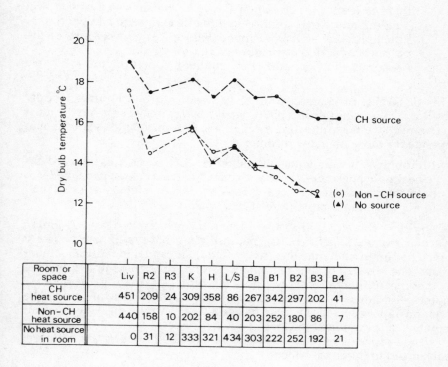

Room or space	Liv	R2	R3	K	H	L/S	Ba	B1	B2	B3	B4
CH heat source	451	209	24	309	358	86	267	342	297	202	41
Non-CH heat source	440	158	10	202	84	40	203	252	180	86	7
No heat source in room	0	31	12	333	321	434	303	222	252	192	21

Figure 7.4 Variation of temperature with type of heating system, UK 1978

Notes: Liv — living room; R — reception room; K — kitchen; H — hall; L/S — landing/stairs; Ba — bathroom; B — bedroom

Source: Hunt and Gidman 1982, p.112, BRE Crown Copyright 1990

The presence of CH continues to be a good indicator of a warm home: the temperature in a CH home in the UK is, on average, about 3°C higher than that in a NCH house (Figure 7.4). In this, as in most temperature surveys, the living room is the warmest room; the kitchen is usually the second warmest, particularly if it contains the CH boiler. Another common feature is that the temperature range in the warmest homes is narrower than in the coldest. In the CH homes, all rooms are fairly warm, about 17-19°C for the 14-hour average. In a NCH home, however, there is generally only one warm room (the living room), each room is colder than those in a CH home and the range within the dwelling is greater: 11.5-17.5°C for the 14-hour means. No survey data have been found of temperatures in homes with the same sort of CH, for instance gas, in relation to income.

There are three reasons why CH homes should be the warmest:

— the heating system is more energy efficient, with lower running costs, than IRH (Table 5.1);
— there is a strong correlation between the presence of CH and building fabric that is more energy efficient, with a lower rate of heat loss because it is younger (Figure 7.3) and because of the presence of added insulation, of all types (Table 4.4);
— the families in CH homes have higher incomes (Figure 7.1; Table 4.4).

This does not, of course, mean that ownership of CH automatically results in a warm home. In a building with a high heat loss, even the most economically efficient heating system, gas CH, may be too expensive to operate for those on a low income:

Although all respondents with either gas CH or gas fires believed gas to be the cheapest fuel, those with gas CH, nonetheless, perceived it to be costly to operate... In 2 cases (out of 19) the systems were switched on for one month only. (Meyel 1987, p.25)

It appears, therefore, that the higher income groups are obtaining warmer homes at least partly because they have purchased energy efficiency improvements: they have used their capital to reduce running costs. The absence of this combination of improvements in low-income homes means that the cost of a unit of warmth is still high, so the poor are purchasing limited quantities of expensive warmth: they are cold. The relationship between the cost of a unit of warmth and the amount purchased is examined in Chapter 9.

A time-series experiment in the 1950s provided households with fuel subsidized to differing extents:

The importance of this experiment was that it demonstrated convincingly that the heating standards observed at that time were not immutable but pegged at their levels by economic constraints. (Fisk 1977, p.396)

The principle would appear to be as true today and can be extended to cover all the factors affecting the cost of a unit of warmth, not just the price of the fuel used. Because the majority of NCH homes are occupied by low-income families, their heating pattern reflects their economic circumstances as well as the higher cost of warmth in their energy inefficient homes. The interdependence of these factors means that present temperatures cannot be used to imply satisfaction with the level of warmth in NCH homes, nor in CH homes occupied by low-income families.

The Hunt and Gidman survey was nationally representative and did not focus on the disadvantaged in any way. Even so, a third of all the rooms or spaces had no source of heat — there was no way in which they could be kept warm except as a result of drifting warmth from other rooms. The temperatures in rooms where there was no source of heat were similar to those with a non-CH source, showing how intermittently IRH are used (Figure 7.4). Thus, the presence of CH indicates a warm home, and, more importantly, the absence of CH means that the home probably has, at the most, one moderately warm room — the living room. The large proportion of rooms and spaces with no heat source confirms the other evidence on defective heating systems in Chapter 5 and demonstrates why many households have no choice but to limit their heating to one room. As stated, the Heating Standard will be difficult to achieve unless there is a heat outlet in all rooms.

Influence of external weather and social characteristics

Interestingly, Hunt and Gidman did not consider the temperatures that they found in low-income homes demonstrated any problems, because the survey was undertaken in a mild March:

> problems with the 'old and cold' ... (and) with other low-income groups may only emerge during cold weather spells. (1982 p.117)

There is little published time-series data on the effect of variations in external temperatures on the levels of warmth recorded in the home. As the weather gets colder, it costs more to heat the home to the same internal temperature, but it is not known which groups in the population maintain the same temperature in the home, regardless of the additional cost, and which groups maintain expenditure at the same level, despite the resultant lower levels of warmth in the home. In their detailed survey of nine households on a Glasgow District Council housing estate, Markus and Nelson found that for seven of the households (and not the same seven that are on the lowest incomes):

> average room temperatures are significantly correlated with outside temperature, more especially in the bedrooms. From this finding it is reasonable to assume that the households are below the income level which would permit fuel expenditure adjustments according to the severity of the weather. (Markus and Nelson 1985, p.5)

Thus, usefully, data on internal temperatures ought to be accompanied by external figures. For instance, the mean outdoor temperature during the Hunt and Gidman survey was 6.6°C which was mild for February-March, but close to the average for the heating season (1982, p.109). They found the coldest homes in the coldest regions (Scotland and the North-East), though only 27 per cent of Scottish homes had CH in 1978 (p.115). The warmest homes, on some measurements, were in the Midlands rather than in the milder South-East and South-West, possibly reflecting the higher proportion with gas CH (Figure 6.1). Greater variations between income groups can be expected during periods of exceptionally severe weather.

> The room gets very cold in January and February. But I don't want to say anything in case he puts the rent up. I can put up with it — you fear worse elsewhere. I can go to the library if it gets really cold. (Elderly private tenant in Smith 1986, p.35)

With data from 40 houses, Pimbert and Fishman found evening living room temperatures were not affected by the external temperature except in uninsulated properties when it was very cold and dropped below 4°C outside (1982, p.6). They conjecture that this may have been due to the inadequacy of the heating system. Even if this is the explanation, it is an indication of deprivation. More data on the relationship between internal temperatures in poor quality housing and external temperatures should emerge from the 1986 English House Condition Survey when the energy study is published. Temperatures were taken in 3,000 homes during November and December 1986 (which were mild) and January 1987, the coldest January on record.

Confirmation of the interrelationship between incomes and energy efficiency comes from the analysis of temperatures by tenure. Half of all local authority tenants are on a low income, but their homes are more energy efficient, with a higher proportion of CH homes (GHS 1987, p.121) and lower rate of heat loss (Table 4.6) than those in the privately rented sector. Therefore, if income levels cannot compensate for energy efficiency variations, privately rented homes should be colder than local authority ones, with owner occupiers the warmest of all. This is exactly what Hunt and Gidman found (1982, p.122), though the living rooms in local authority housing are slightly warmer during the day than those of owner occupiers — a result that shows that the heating standard is soundly based on occupancy levels.

Many low-income households such as pensioners, the unemployed, housewives with young children, are in the home for most of the day. The temperature in the homes of these groups should be the warmest, because of their high occupancy rates. The group that is most studied is the elderly and in a survey across all retired people in 1972, before the first oil crisis, reported by Wicks (1978, pp.42-4) average temperatures were:

the living room, morning — 15.8°C
the living room, evening — 18.1°C
the bedroom, 24 hours — 13.9°C

All these are far below the DHSS recommendation of 21°C for the whole dwelling that was current then (Table 6.3). The other set of comprehensive data on elderly people is in the Hunt and Gidman survey, which showed that the over 65s, like 18-24 year olds, are likely to have low temperatures but not CH (1982, p.120). The temperatures recorded, for instance 15.5°C MIT over 14 hours, were not as low as Wicks describes. Whether this is because of different samples or the six-year gap between the surveys is not known. Whilst advice from the DHSS, RIBA and the Institute for Housing (Table 6.3) is that the elderly should live in warmer homes than younger people, the opposite appears to be true in practice. This confirms that the elderly are inclined to restrict their heating to dangerously low levels, particularly if on a low income.

Although much of the evidence is fairly old, temperatures have been rising and are probably still below predicted comfort levels at all times except, perhaps, living rooms in the evening. Increased warmth is, therefore, a latent demand, that will be met as increased energy efficiency, or increased income, permit. Confirmation of this comes from the data for other countries given below, which show higher internal temperatures than are commonly found in the UK.

Excess winter deaths

Because of the shortage of data, the extent of cold housing conditions has to be gauged partly in terms of excess winter deaths (EWD): all age-groups, except 5-24 year olds, suffer more deaths in winter than in summer (Curwen 1981, p.18). If winter in Britain is defined as the months December-March inclusive, the number of people who die in these four months exceeds the number that die in an average four months in the rest of the year to give the number of EWD (see Glossary). The relationship between low temperatures indoors and deteriorating ill health is given in Chapter 6. Whilst there may be uncertainty about the exact temperatures and periods of exposure that trigger these diseases, the evidence of a causal link between being cold and morbidity or mortality is strong:

- chilling for six hours results in a thickening of the blood and an increased risk of heart attacks and strokes, and mortality from these two causes in Britain increases linearly as air temperature falls from summer to winter (Keatinge et al. 1984, pp.1405-8);
- deaths from respiratory and circulatory disorders, including heart disease, are strongly correlated with poor space-heating (Wynn and Wynn 1979, p.112);
- in England and Wales, social class V babies (<1 year) are nearly five times more likely to die from bronchitis or pneumonia than babies in social class I (see social classes in Glossary). These two diseases thrive in the cold; even if the body is warm, simply breathing cold air accelerates the progress of respiratory illness (Wynn and Wynn 1979, pp.208-10);

— to remain healthy, the body must maintain a fairly constant temperature of 37°C (98.6°F). If the inner body temperature falls to 35°C (95°F), the individual is suffering from hypothermia — the cause of death for up to 600 people in the UK in the 1986/7 winter (Hansard 24 July 1987, WA col 692), which represents about 2 per cent of EWD. Deaths from hypothermia were an early indicator of fuel poverty (Chapter 2), though they constitute a small proportion of EWD.

Keatinge compared deaths amongst elderly people in warm, sheltered housing with those amongst the over-65s in the general population (1986). He found that overall mortality was lower among the housing association residents, although they are a population selected because of increasing infirmity to live in accommodation with a warden. However,

the continuous high daytime temperatures did not prevent mortality among the residents from rising in winter by a percentage similar to that among the general population (Keatinge 1986, pp.732-3),

perhaps because of the residents' preference for open windows at night. The beneficial effect of warm homes was apparent though it is not clear why there was not a greater decline in EWD, as would have been anticipated in these warm homes. The methodology involved large-scale data on deaths, which was combined with temperature surveys on 14 residents and their homes. The design of the research may be a reason for uncertainty about the findings, though there are obvious problems in relating deaths to temperatures whilst living.

Keatinge believes that the additional winter deaths resulted from getting cold outside, typically whilst waiting at bus stops. As Collins has commented this:

does not explain why such winter excess mortality apparently fails to occur in countries with much colder outdoor winter conditions. (1989, p.259)

The debate continues as to which causes of death are most associated with temperatures indoors or outdoors, particularly for the young and the old. In Northern Ireland, there are 49 per cent more deaths from heart attacks and 53 per cent more deaths from strokes in January than in August (McKee 1990, p20). The numbers of sudden infant death syndrome (SIDS) cases show a direct relation with colder weather (Murphy and Campbell 1987, p.63). This study confirmed the importance of indoor temperatures in England and Wales (p.70), as most of these are known as 'cot deaths'.

Britain has one of the highest infant mortality rates (babies under one year) among the advanced countries: 60 per cent and 80 per cent above Sweden and Japan respectively in 1986 (HC 54 1988, p.34). In the same year, the infant mortality rate in England and Wales increased for the first time since 1970 (OPCS Monitor 15 Dec. 1987, p.2). February 1986 was particularly cold (-0.4°C on average) and this appears to have been a contributory factor. As the Social Services Select Committee report:

research from Sheffield linking infant deaths from treatable disease and minor illness to levels of unemployment underline the importance of the level of parental income to the survival of vulnerable babies, in that the poorest are likely to be least able to counter exceptionally low temperatures. (HC 54 1988, p.xx)

Further research into SIDS in February 1986 showed the increase was:

higher than expected in the areas with the poorest population and the greatest dependence on electricity for domestic heating. (Bentham 1990, p.5)

There is no differentiation between types of electric heating, though, as shown in Chapter 5, only night storage heaters are relatively inexpensive. In Sweden, with a low infant mortality rate, the advice is to keep babies at temperatures above 20°C, or higher if the child is sick. Wynn and Wynn went on to state:

Much stress is often placed on overcrowding as a cause of ill health, but space heating appears to be much more important in the prevention of somatic illness. There are many overcrowded homes in Finland but they are warm at all times and there is no increase in infant deaths from respiratory illness in winter. (1979, p.210)

Thus, there is good evidence that exposure to cold results in increased health risks, particularly for babies in the first year of life and people aged 75 or over, who show the greatest rise in mortality in winter. Excess winter deaths double for every 9 years of advancing age over 40 (Collins 1989, p.258). Statistics referring to the general population, therefore, disguise the influence that the factors examined could be having on those most at risk.

In England and Wales, comparing the summer quarter with the winter quarter, there is a 20-47 per cent increase in deaths in winter, and twice as many people die in a cold January as in a warm August. In the last twenty years, the greatest number of EWD occurred in 1976, when there were 59,000 additional deaths, and the least in 1987 when the number was 25,000. For the UK, the annual variation in EWD is, approximately, in the range 30-60,000.

The OPCS have established (Curwen and Devis 1988, p.17-20) that since 1949/50, the number of EWD in England and Wales is related to influenza epidemics, the weather and is gradually reducing over time at rate of 530 deaths a year. The weather variable means that for every 1°C that the temperature is colder than average over the four months December-March there will be 8,000 extra deaths; the converse is also true. The close relationship between external weather and the number of excess winter deaths, for the years 1950-1 to 1977-8, is shown by Curwen (1981, p.14) and updated to 1984-5 by Boardman (1986b, p.3). These demonstrate the importance of the external weather conditions and the (slow) rate of improvement in the time-related factor. However, even at this slow rate, the cumulative impact of the decrease in EWD means that, as a result of the decline since 1949/50, there were about 20,000 less deaths in winter in 1985/6; to reduce the base figure to 0 would take another 60 years at historic rates. Because of the decline in the base figure that has already taken place, the additional deaths caused by severe weather represent an even greater proportion of the total.

The period over which the temperature fall is measured appears to be important. McKee found no significant association with annual temperature range and minimum average monthly temperature explained 46 per cent of the variation in deaths (1989, p.179). With SIDS, the speed and the severity of the fall four to six days earlier was found to be crucial:

> the peak of SIDS deaths occurs close to the time when the weather is deteriorating most dramatically. (Murphy and Campbell 1987, p.68)

There is, therefore, clear evidence that cold weather causes more extra deaths in Britain, with most evidence pointing to the effect of cold living conditions rather than external exposure.

There are two other important findings:

> a marked social gradient in excess winter mortality ... the percentage of EWD being almost exactly twice as great in Social Class V as in Social Class I (Curwen 1981, p.18); and

> a bad winter does not bring forward deaths which would have otherwise occurred shortly after the period of high risk. (Curwen and Devis 1988, p.18)

These EWD are premature deaths, that can be avoided, and they occur disproportionately amongst the poor. The strong social class bias in ill health has been identified most recently by the British Medical Association (1987) and the Health Education Authority (1987). Both recognize the pervasive influence of bad housing conditions.

International comparisons

The range in the number of monthly deaths throughout the year (as measured by the coefficient of variation) is greater for the UK and Ireland than for countries with colder or similar external climates such as Canada, Denmark, Norway and France (Table 7.2). Among these countries, the British Isles have both the warmest winters and greatest seasonal variations in mortality. A coefficient of variation of 0.10 is twice as great as one of 0.05. The number of excess winter deaths is influenced by annual variations in the severity of the winter, or other factors, but for most countries there is an improvement in the coefficient of variation from 1978-84. Thus, the external temperature is not the explanation for variations in EWD between countries, although it does contribute to the annual swings within England and Wales. Alderson (1985) reviewed papers studying the effect of other factors that vary with the seasons, such as humidity, rainfall and air pollution, and although the latter may be a contributory factor, external temperature has the greatest impact.

Table 7.2 Seasonal mortality and external temperatures: international comparisons

	Mean external temperature in January	Seasonal mortality coefficient of variation	
	°C	1978	1984
Ireland	5.0	0.15	0.13
Northern Ireland	4.5	0.12	0.14
England and Wales	4.1	0.13	0.10
Scotland	3.7	0.12	0.11
France	3.3	0.07	0.06
Denmark	0	0.07	0.06
Norway	−1.1	0.05	0.04
Austria	−2.7	0.11	0.05
Sweden	−2.7	0.07	0.05
Finland	−3.0	0.05	0.04
Canada	−7.8	0.07	0.06

Sources: Col 1: Based on Kendrew 1961, pp.378-83, 458. Average temperatures do not vary significantly; Cols 2 and 3: Derived from UN Demographic Yearbooks 1980 and 1985, Table 30

Therefore, in other countries, the effect of cold weather on people's health is reduced considerably, but in the British Isles the impact of cold weather is still significant. For instance, deaths of neonatal babies, aged 1-12 months, show a marked seasonal swing because of respiratory illness in the UK, although:

> in Finland, the Netherlands and Sweden there is now no seasonal swing in infant deaths in spite of more severe and longer lasting winters. (Wynn and Wynn 1979, p.208)

The evidence earlier in this chapter showed that many British families have indoor temperatures that are inadequate for maintaining health, with rooms well below the comfort zone, even during the day. Similarly, due to economic constraints, it appears that low-income families often cannot afford to increase expenditure, and maintain the level of warmth, during periods of severe weather. Thus, there is a closer relationship between external and internal temperatures in the UK than in many other countries.

There is very little information on domestic temperatures in different countries, but a few comparisons can be made (Figure 7.5). Not all the data are collected on a comparable basis: for instance, temperatures given for Scotland, England and Wales result from measurements taken when the interviewer had access to the home in the day and evening — a

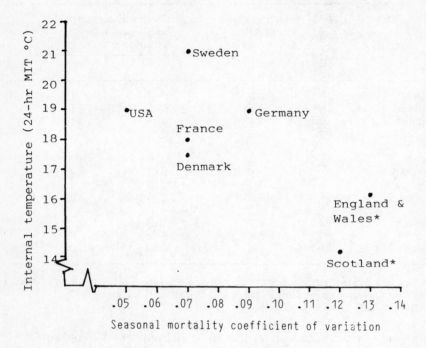

Figure 7.5 Seasonal mortality and winter temperatures inside the home: international comparisons, 1978

Note: *14-hour mean internal temperature

Sources: Hunt and Gidman 1982, p.116; Schipper *et al*. 1985, p.3; UN 1980, Table 30

14-hour period. If it were possible to give a 24-hour mean, to compare with other countries, the temperature would be lower still, probably by at least 2°C. There is a strong negative correlation (r = -0.73; p<0.05) between the variations in seasonal mortality and internal temperatures given in Figure 7.5, showing that EWD are significantly linked to cold homes.

Within Britain, indoor temperatures vary significantly with aspects of energy efficiency, such as the date of the building, and thus heat loss rate, or the presence of CH. The 'ever-wider use of CH' is cited by Sakamoto-Momiyama as the reason for the decline in the number of EWD in the USA (1977, p.21). Keatinge *et al*. believe that the increasing ownership of CH in Britain has resulted in warmer homes and therefore that as:

excess coronary and cerebrovascular mortalities in winter did not fall
significantly from 1964 to 1984 ... (the findings) strengthen these
indications that outdoor excursions rather than cold houses are the
main cause of the arterial deaths that now cause most of the excess
mortality in winter. (1989, p.76)

The alternative view is that 'those most vulnerable to cold conditions —
the elderly, very young, and the infirm — are not benefiting from general
improvements to the level of warmth in the home' (Boardman 1986b). The
debate continues.

Examples of the effect of energy efficiency factors on seasonal mortality
have cited a wide range of countries:

—   the widespread availability of low-cost geothermal energy as the
    reason for the absence of seasonal variation in Iceland, despite an
    average winter temperature below freezing (McKee 1989, p.180);
—   improved housing conditions in Finland over the last 50 years
    resulting in a substantial fall in EWD (Nayha 1984);
—   abundant low-cost natural gas in the Netherlands as the
    explanation for a lower seasonal variation than neighbouring
    Belgium (McKee, letters, *The Lancet*, 2 Sept. 1989, p.564);
—   the close association between seasonal variation and latitude in
    Chile, where there is little insulation or central heating (ibid,
    p.565);
—   EWD in New Zealand are high and rising, leading to a call for the
    continuation of the requirement for all dwellings to meet
    minimum levels of thermal insulation (Isaacs and Donn 1990, in
    press).

There is, therefore, widespread agreement that winter mortality levels
will decline if the energy efficiency of the housing stock is improved.

International comparisons of the energy efficiency of new homes can be
made using the Building Regulation standards from different countries
for a standard home occupied by an average family (Figure 7.6). The
Milton Keynes Energy Cost Index (MKECI) gives a measure of the cost of
all energy used in a home, per square metre of floor area. Fuel bills in a
new home in 1985 in the UK would be 70 per cent higher than in a house,
in the same location and of the same basic dimensions, but built to the
standard of the Finnish building regulations (roughly 173.0 ÷ 99.1) and
providing the same level of comfort to the occupants. In 1985 the British
Building Regulations required less insulation than those of any of the
other countries shown in Figure 7.8 and that is why the faster rate of heat
loss would result in larger fuel bills to obtain the same standard of
warmth. The 1990 revision would bring the British MKECI down to about
150-160, but other countries have also upgraded their standards as well.
Owing to the success of the Milton Keynes Energy Park, new buildings
now have to meet a target of MKECI 100 to be built there.

The likelihood is that similar discrepancies existed in the past and

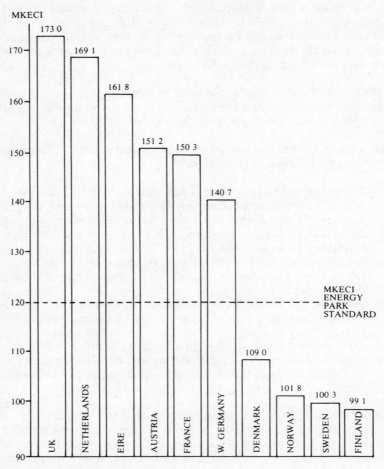

MKECI FIGURES FOR HOUSES BUILT TO THE CURRENT BUILDING STANDARDS OF VARIOUS COUNTRIES.

Figure 7.6 Comparison of Building Regulation standards in ten European countries using the Milton Keynes Energy Cost Index (MKECI)

Note: Comparison based on Pennyland three bedroom, five person, single aspect, end of terrace house. For a description of MKECI see Glossary

Source: Evidence given to the Select Committee on Energy by Potter (HC 87 1985, p.47)

therefore that our present housing stock is similarly inefficient in comparison with older European stock. For instance, double glazing has been required in Sweden since the 1930s, and quadruple glazing is now mandatory, whereas single glazing is still acceptable in the 1990 British

Building Regulations (Chapter 4). Thus, the whole UK housing stock is probably less energy efficient than that of most other European countries and the cost of keeping warm consequently higher.

In the absence of more definite data, it is suggested that there is a causal link between the low level of energy efficiency in British dwellings, cold homes and the high seasonal mortality rate. As all three components are individually correlated with low incomes, EWD indicate the existence of fuel poverty and the hardship and distress caused by cold homes:

> I can't wash in the bathroom in the winter at all. It's so cold it wouldn't be safe. (Nottingham Heating Project, 1985b)

Government policies and cold weather

The expected variation in temperature (based on 20-year averages), across the whole of the UK, is not reflected in assistance to the poor as benefits are paid at the same rate in all regions. In 1984 Gordon Wilson, MP, unsuccessfully introduced a Private Members' Bill to zone the country into four regions, for benefit purposes, with a base figure rising by 10 per cent, 20 per cent and 30 per cent to reflect the increased costs of domestic heating in a normal year (Hansard 25 Jan. 1984). This would have been a simple and appropriate way to reduce the hardship created by predictable climatic factors: approximately a third of the population is in the base zone, half in the 10 per cent region and the remaining 17 per cent in the highest two.

Periods of unusually cold weather, such as February 1986 and January 1987, create a great deal of hardship and political furore. The problem of trying to keep warm, particularly for the elderly, is the subject of extensive media coverage and parliamentary debate. IS recipients are entitled to additional money during periods of exceptionally severe weather (ESW), subject to eligibility. The DSS has defined ESW in several different ways since the introduction of the scheme in 1981, with the present scheme being introduced early in 1989. This enables IS claimants with anyone under 5, over 60 or chronically sick or disabled in the household to claim, if the average temperature for seven consecutive days is less than 0°C. Only claimants with savings of less than £1,000 are entitled to apply and there is a standard payment (irrespective of actual use) of £5 for each week of ESW. It is estimated that about 1.4m claimants are eligible on this basis (Hansard 6 Feb. 1987, WA col 874) — 26 per cent of all SB recipients at that time. As actual expenditure to cover the particularly severe January-February 1987 totalled £10.2m (Hansard 2 Nov. 1987, WA col 557), each eligible claimant received an average of £7.30, as the extra payment for the whole winter. Thus, assistance is provided on a severely restricted basis. The administrative costs of the scheme are extremely high, and amounted to an additional £5m for the period January-March 1987 (Hansard 2 April 1987, WA col 622-3). Each £1 of benefit cost 50p to pay out.

The present definition is based upon an assumption that it would be appropriate to make ESW payments once in every two years. By implication, therefore, exceptional means one week out of 78 (if the heating season is the 39 weeks from October to May). Many people will agree with Fairmaner's view that this is 'too harsh' (1988, p.12). His proposal is that the threshold should be '5 days with the maximum daily temperature one standard deviation below the daily mean maxima' (p.12). Whilst more generous, this would be difficult for the public and the media to interpret and wide publicity is needed to ensure that the benefit is claimed.

The ESW policy illustrates a further problem in relation to cold weather — the fact that it has been triggered is announced retrospectively. To keep warm, people need to know that the additional money is available whilst the weather is cold, without having to wait to see if it lasts for seven days. These administrative problems indicate the difficulties of devising a policy to lessen the impact of unpredictable cold spells. The lack of policies to reduce the impact of predictable levels of cold, for instance based on average DD, is even more surprising. For the immediate future, periods of ESW will continue to create considerable hardship for people on a low income in energy inefficient homes, where the cost of each additional degree of warmth needed to compensate for the cold outside is high.

> Today has been very cold. I can feel the drop in temperature. I have also run out of coal, so I will have no coal tomorrow unless I can borrow the money to buy a small bag. (A single mother with four children, quoted in Stitt 1990, p.14)

Conclusions

There is a considerable variation in temperatures within UK houses and the most reliable predictors of warm homes are:

—   an economically efficient heating system, such as CH;
—   a well insulated building, for instance a modern dwelling; and
—   occupants with above average income.

The highest temperatures are in owner occupied, energy efficient homes, where 24-hour, whole house MIT of 18-19°C are recorded. The heating standard suggested in the previous chapter results in a 24-hour whole house MIT of 18°C. This standard appears to reflect a reasonable aspiration for a low-income family in the home most of the day, both in relation to the minimum necessary for health and comfort and the standards achieved by other households.

The average British home is cold in comparison with other advanced countries. The lowest temperatures are found in homes with no central heating with a MIT of no more than 14°C — even this temperature assumes a temperature rise since 1978 that is not supported by research. There is no evidence of temperature trends by income group or tenure, although CH homes are becoming warmer, probably more quickly than

NCH. If this is the case, the gap between the warmest and coldest is increasing, and fuel poverty is worsening, on one of the most important measurements.

Cold homes cause ill health for the occupants. The number of excess winter deaths is declining, but only slowly, and the impact of cold weather appears to be just as great as in 1949/50 with 8,000 EWD/°C of additional winter cold. This may indicate that those most at risk during cold weather, for instance the elderly, are not noticeably benefiting from better methods of heating and other improvements in energy efficiency and, therefore, may not have warmer homes. This is entirely in keeping with the evidence in earlier chapters that energy efficiency improvements have occurred mainly in the homes of better-off households, who are less vulnerable to the effects of cold weather.

The confusing role of the external climate — it significantly affects the number of deaths in winter in this country, but not in others — implies that the crucial factor is the way in which the cold weather is modified by the housing stock. Seasonal mortality is less pronounced in countries with warmer homes, but a colder external winter climate. The implication is that the high rate of winter deaths in Britain is caused by cold homes and thus many of these deaths are avoidable and preventable. To put this in context, if seasonal mortality in Britain could be reduced to the level of Scandinavian countries, EWD would be halved to bring the annual total down to 15,000-30,000: three to six times the number of road deaths (5,500 in 1987 — *Social Trends* 1989, p125) could be saved through warmer homes.

The evidence on energy efficiency levels, temperatures in the home and winter deaths all demonstrate (with varying levels of certainty) a social class or income level bias. So the prevention of EWD will depend largely upon policies to improve the energy efficiency of low-income homes. Therefore, a recognition is needed, in building and environmental health standards, that a home must be warm to be healthy. Good health may depend on adequate warmth, just as much as it does on freedom from damp, and sufficient air and light. The elderly, the ill and babies are some of those most vulnerable to the effects of cold. These groups do not (necessarily) need higher temperatures *per se*, they are just more affected by a lack of adequate warmth than healthier members of the community. The need is to make sure that they have energy efficient homes that provide them with affordable warmth.

8 Energy purchases

The amount of energy purchased by low-income households (and others) completes the evidence section. The trends from previous chapters on the energy efficiency of the building fabric and changing ownership of heating systems, when combined with data on energy consumption, should clarify the evidence on changing levels of warmth. Thus, expenditure information may help to confirm whether fuel poverty is or is not worsening and provide further evidence about the effects of fuel substitution, particularly for low-income households.

Total domestic energy consumption in the UK is used to validate trends in individual households. There are considerable statistical complexities in trying to dovetail results as varied as total sales of fuel to the domestic sector, average expenditure on energy by families, measured temperatures extrapolated from a 1978 survey and estimated heat loss. The benefit of including energy consumption is that it is a robust, consistent set of statistics with which to anchor other data. The size of some of the adjustments required, together with the shortage of data in other areas, demonstrates both the problems and the lack of government interest in modelling the situation.

The analysis works progressively from total domestic consumption, through delivered energy purchases per household, to useful energy per household. This is then compared with average household expenditure data and, finally, the latter is subdivided by income. The DEn's statistics on total delivered energy are assumed to be the most reliable of the sources being used in this chapter, because the data come from (a relatively few) suppliers. They are thus examined first.

Average household trends

Delivered energy

In order to examine national consumption trends, the effect of external weather has to be controlled, because of the increase in the demand for space heating when it is cold. The official method of temperature correction is for primary energy consumption only, covering all uses (DUKES 1987, p.5) and there is, unfortunately, no recommended method of standardizing delivered domestic energy consumption for the effect of the weather. Some work has been undertaken employing climatic correction factors based on DD figures (Leach and Pellew 1982, pp.8-9; Chateau and Lapillone 1982, p.22), though neither source explains the method used. Leach and Pellew obtained a figure of 6 kWh/household/ DD when winter energy consumption is plotted against DD, 1974-81 (1982, p20), which is approximately 166 PJ pa/°C. The BRE estimate that for every extra 100 DD domestic energy consumption increases by 45 PJ (Baldwin *et al.* 1986, p.6). If the mean temperature, over the whole year, is one degree colder (approximately to 365 DD), the predicted increase would be 164 PJ of delivered energy, identical to Leach and Pellew.

The BRE regression line has been added to actual consumption plotted against annual variations in mean temperature (Figure 8.1). The fit looks relatively good except for 1972 and 1986-8. The temperature for the whole

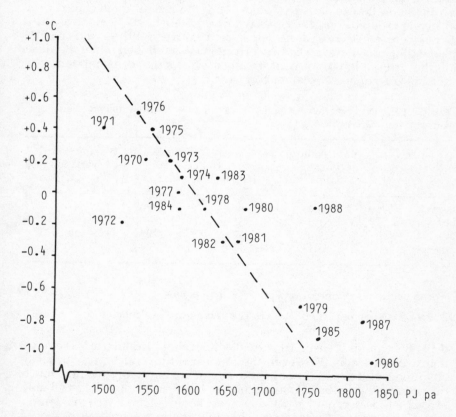

Figure 8.1 Total domestic delivered energy consumption in relation to the annual variation in mean external temperature, UK 1970-88

Note: The analysis is based on the average temperature for 1951-80. For an explanation of the regression line, see text.

Source: Based on DUKES

calendar year has been taken, rather than the winter period, in order to match total fuel consumption, though this introduces further problems. For instance, 1972 was cold because temperatures in the summer, not the winter, were below average and that is why fuel usage did not increase as usual. Consumption was higher than would have been predicted in the last three years plotted, though only 1986 and 1987 were unusually cold. As 10 of the 11 years 1978-88 have been colder than average, any analysis of recent energy consumption and expenditure, that does not correct for the weather, will overestimate the level of warmth being attained. The BRE correction figure is used in the subsequent analysis, despite its now suspect value.

The number of households in the UK has been increasing steadily during this period from 18.7m in 1970 to 21.9m in 1988, with a concurrent reduction in household size (Table 8.1). If the results are averaged over five-year groups, household energy consumption has either increased by 3 per cent or decreased by 5 per cent between 1970-4 and 1985-8, depending on whether weather correction is undertaken or not. Because of the crucial importance of weather on energy consumption, the remaining figures are weather corrected.

Table 8.1 Effect of weather correction on average household delivered energy consumption, UK 1970-88

Years	Not weather-corrected (GJ pa)	Weather-corrected (GJ pa)	Household size (people)	Temperature variation* (°C)
1970-4	80.8	82.0	2.88	+0.1
1975-9	81.0	81.2	2.75	0
1980-4	79.0	77.9	2.68	-0.1
1985-8	83.1	77.6	2.55	-0.7

Note:
*For the period, in comparison with 20-year averages

Sources: Based on data from DUKES, FES and author's calculations

If all other things were equal, a smaller household would use less energy than a larger family. However, the amount of the reduction is not easily quantified, as it depends upon the behaviour of the family and how much they combine activities. At a minimum, the smaller household will only use less hot water, if other activities were largely shared. At a maximum, with independent individuals, halving the household size would approximately halve energy consumption. There has been an 11 per cent reduction in household size compared with a 5 per cent drop in energy consumption (Table 8.1) indicating either that the economies of scale with large families are minimal, or that they are being offset by a demand for higher energy services, or both.

Useful energy

The focus of this book is on the demand for energy and the way that it is used. The amount of useful energy obtained from a given quantity of delivered energy, as shown in Chapter 5, depends on the efficiency of the appliance. To convert total UK delivered energy into useful energy requires the use of conversion factors that summarize the technical efficiency of the different appliances in the home that use that fuel. Estimating average technical efficiencies across the whole range of appliances in the housing stock for the range of uses involves considerable guesswork. One of the factors is the proportion of each fuel used under the main domestic categories (Table 8.2). For solid fuel, gas and oil the predominant use is heating, whereas most electricity is consumed in appliances.

Table 8.2 Delivered energy consumption in the domestic sector by use and fuel, UK 1986 (%)

	Solid fuel	Gas	Oil	Electricity	Total
Space heating	11	41	5	4	61
Water heating	5	15	1	3	24
Cooking	-	3	-	2	5
Appliances	-	-	-	10	10
Total	17	59	6	18	100

Source: CEGB (1988)

Based on several sources, Table 8.3 gives the conversion factors used here, for instance to derive the data in Figure 8.2. As demonstrated in Chapter 5, the increase in efficiencies over time comes only partially from technical improvements to the equipment, for instance gas. The greater influence is the substitution of one heating method (and fuel) for another. Hence, the greater use of stoves and enclosed room heaters results in the efficiency with which solid fuel is used. The changing proportions are demonstrated by the weighted average: fuel in the domestic sector is being used a third more efficiently in 1988 than it was in 1970, largely as a result of fuel substitution.

Because of the lack of agreement between the different sources, the resultant data on useful energy has to be treated as indicative, rather than accurate in detail. However, all sources agree that gas is used more efficiently than solid fuel and this is the most important comparison. It is the substantial increase in the use of gas, particularly as a substitute for solid fuel, that has had most influence on the useful energy trend shown in Figure 8.2. By 1988, 60 per cent of all useful energy came from gas in the

Table 8.3 Average technical efficiencies for the stock of domestic energy appliances 1970-88 (%)

	1970	1988
Solid fuel	30	36
Gas	55	67
Electricity	90	90
Other, mainly petroleum products	80	80
Weighted average	51	68

Sources: used include: NEDO 1974, p.28; DEn 1978b, p.64; Leach and Pellew 1982, p.8; Pezzey 1984, p.161; Henderson and Shorrock 1989, pp.20-1

average household. This 33 GJ is three times as much as the useful energy from gas in 1970 (11 GJ). Household electricity usage is at exactly the same level, whereas use of petroleum products has declined slightly.

The net effect was an increase in the amount of useful energy obtained

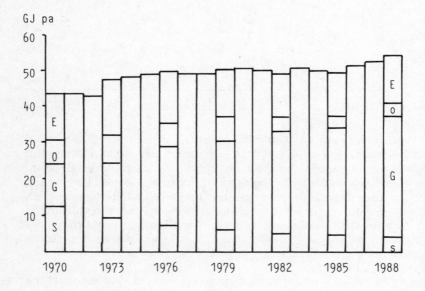

Figure 8.2 Weather-corrected useful energy per household, UK 1970-88

Notes: E — Electricity, O — other, mainly petroleum products; G — gas; S — solid fuel

Sources: Author's calculation, based on DUKES

by the average household between 1970-5, and since then useful consumption has been fairly stable at 49-50 GJ/household pa until 1985. The increase in the last three years of the data occurred as a result of a rise in gas purchases, when weather corrected, whilst consumption of other fuels stayed stable.

The amount of fuel used for heating, as a proportion of either consumption or expenditure, varies considerably between households: it is definitely lowest in energy efficient homes and probably highest in low-income ones. However, as all energy use contributes to incidental gains within the house, total fuel consumption is an acceptable proxy for spending on heating alone and avoids the need to add incidental gains back in. These figures for useful energy per household, therefore, represent the amount of purchased energy contributing towards warmth in the dwelling; other incidental gains are not included.

Expenditure on useful energy

This is where the calculations get very messy. There are only two published figures for general UK household energy expenditure. Using 1988 as an example, these were:

Average weekly household expenditure on fuel (FES, p.15)   £10.48
Total domestic energy expenditure (DUKES, p.25)   £11,005m

For both these figures to be correct, there would have to be 20.19m households in the UK in 1988. The number of UK households is not a published statistic, except for census years. The DOE estimate that there were 21.9m households in the UK in 1988 — 8.5 per cent higher than 20.19m. Therefore, there is a 8.5 per cent difference between the two expenditure figures quoted above, with no way of establishing why households are reporting so much more expenditure than the industries. Unfortunately, in 1970, the base year, the discrepancy was even higher at 15 per cent, again with households reporting more expenditure than the industries' sales. It is indicative of government interest in the domestic sector that no analysis of these differences appears to have been undertaken. The net effect for this exercise is to introduce yet another note of caution about the findings: any trend from 1970-88 based on FES data will vary by nearly 7 per cent (15-8 per cent) from exactly the same trend using information from DUKES. Furthermore, it is extremely difficult to use detailed expenditure data from FES in conjunction with consumption information from DUKES.

Using DUKES data on expenditure in comparison with consumption figures from the same source, the next stage in the analysis is to consider household expenditure on useful energy. Annual expenditure on energy by individual households, when weather corrected, has risen from £82.50 in 1970 to £498 in 1988 — approximately the same proportion of total expenditure (5-6 per cent) throughout. Low-income households have been consistent as well, spending 10-11 per cent of their expenditure on fuel (Figure 3.2 and Table 3.4 for 1988 FES data). In both groups, the weekly budget on all items has risen in real terms, so these fixed proportions appear to represent an increase in real expenditure on energy. This is the point made by Dilnot and Helm that a constant proportion of rising real expenditure demonstrates a demand for a higher level of warmth:

> as the general standard of living has risen in the UK, the required amount of heating and lighting has also risen. (1987, p.37)

When weather-corrected fuel expenditure is considered in money terms, and not as a proportion of income, then the real value has fluctuated considerably (Table 8.4): people are probably spending more on fuel in real terms, but the trend is uncertain. (A cautious note is necessary, because the analysis is dependent upon the accuracy of the weather correction method used.) In fact, as fuel prices have increased faster than the rate of inflation, the quantity of delivered energy purchased by the average household has declined, as confirmed in Table 8.1. Thus, it is not possible to endorse Dilnot and Helm's statement, at least for the years 1970 onwards, that a constant proportion of rising real expenditure, in itself, demonstrates a demand for a higher level of warmth.

Table 8.4 Real cost of useful energy per household, weather-corrected, UK 1970-88 (per annum)

	Total energy expenditure (1988£)	Useful energy purchased (GJ)	Cost per unit of useful energy (1988£/GJ)	% of useful energy from gas
1970	472	43.4	10.87	26
1973	457	47.4	9.64	32
1976	516	49.6	10.40	42
1979	476	50.2	9.49	50
1982	542	49.0	11.06	56
1985	519	49.2	10.55	58
1988	498	54.2	9.19	61

Note: The same weather-correction factor has been applied to both energy consumption and energy expenditure.

Source: Based on DUKES and previous data

The uncertain trend in real fuel expenditure combines with the slightly increasing useful energy purchases (Figure 8.2) to give an erratic sequence of figures for the cost per unit of useful energy in Table 8.4. It is, thus, not possible to state that the cost of a unit of useful energy has consistently dropped in real terms. This is not entirely surprising, because:

— the weather correction procedure for both consumption and expenditure is a considerable simplification of a number of factors and appears to be less accurate for the years 1986-8;

— fuel price movements have varied considerably. As Table 8.4 shows, an increasing proportion of useful energy is coming from gas and the real price of gas was 30 per cent lower in 1979 than in 1983 (Figure 2.2).

A feasible explanation might be that the average household has tried to keep real expenditure on fuel constant, despite substantial changes in the price of fuel and weather, whilst at the same time increasing the level of service obtained through energy efficiency improvements.

One of the conundrums with domestic energy expenditure is why it should be so stable (as a proportion of income), when, as shown in Chapter 7, houses are quite cool and additional purchases could be expected. This is especially curious in the average household, where energy is ranked 8 out of 10 in expenditure priorities, only slightly more important than alcohol. Nothing appears to be known about the reasons for household priorities and whether fuel expenditure is driven by the level of service obtained, is psychologically fixed at this proportion of income, or why more money is not spent on fuel. This uncertainty is compounded by the likelihood that few individuals can accurately price a fuel, although they may be aware of the size of their total fuel bills.

Much of the increase in useful energy has been achieved through the substitution of gas for other, less efficient, fuels such as solid fuel, though this is also slowing down (Table 8.4). The purchase of useful energy is only one of the factors influencing the level of warmth in the home. In addition to the increase in useful energy, there have been improvements in the insulation of the building fabric. This is discussed further below.

Energy expenditure by income

The FES is the only source of information on energy expenditure that disaggregates by income, using a representative sample of 7,000 households. No data are available on actual consumption of energy by income group. Thus, information on energy use in poor households can be obtained only through converting FES expenditure data into consumption figures, through average prices and average efficiencies. The conversion of expenditure data into consumption estimates is, therefore, an imperfect process, giving useful results rather than statistically-certain findings.

Low-income households in 1988 are still not purchasing as much useful energy as other households were in 1970. By implication, there is at least a 20-year time-lag between the two groups. As the same technical efficiency factors have been applied to both income groups (as given in Table 8.3), although fuel substitution and improvements in the efficiency of conversion have occurred disproportionately in the higher income

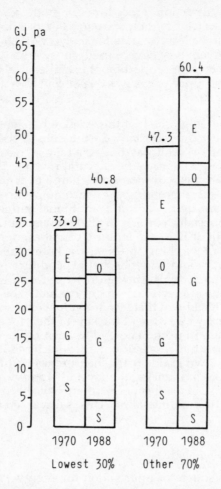

Figure 8.3 Weather-corrected useful energy, per household, by income group UK 1970-88

Notes: E — Electricity; O — other, mainly petroleum products; G — gas; S — solid fuel

Sources: Based on FES and DUKES; fuel prices from the Electricity Council (pers comm), British Gas Corporation (pers comm), and Sutherland (per comms)

groups (Chapter 5), the useful energy purchases by low-income households may be slightly overestimated in Figure 8.3.

The same data are summarized in Table 8.5. As real expenditure has stayed the same for each group, the divergence in standards is due to useful energy purchases increasing at a faster rate in better-off families than in a low-income one. Thus, not only are the poor deprived, but the gap is widening. All of the additional consumption in the richer households comes from gas, whereas both gas and electricity usage have grown in low-income households. The reason for the widening gap in useful energy purchases is that better-off households have achieved a faster rate of fuel substitution than poor ones. This is a finding that confirms the conclusions from Chapter 5. For instance, as a proportion of useful energy purchases, gas has risen from 26 to 62 per cent in better-off homes, but from 24 to 52 per cent for the poor. The standard of living, as measured by useful energy consumed per person, is increasing 50 per cent faster in richer families than in poor ones.

Table 8.5 Changes in the cost of useful energy for low-income and other families, weather-corrected, UK 1970-88 (per annum)

	30% lowest incomes			70% other		
	1970	1988	% change	1970	1988	% change
GJ/household	33.9	40.8	+20	47.3	60.4	+28
Real expenditure (1988£)	438	437	0	587	584	0
Cost (1988£/GJ)	12.92	10.70	-21	12.41	9.67	-28
GJ/ person	17.8	23.9	+34	13.9	21.1	+52

Note: The expenditure includes the cost of standing charges.

Sources: As for Figure 8.3

Each gigajoule of useful energy costs low-income families more than other households: the poor are buying more expensive useful energy in 1988, just as they were in 1970 (Table 8.5). Furthermore, this differential is widening as the real cost of useful energy has dropped by less for low-income families (-21 per cent) than for other households (-28 per cent). In 1988, the poor had to pay 11 per cent (£10.70 ÷ £9.67) more for a unit of useful energy than other households. In order to determine the differential in the cost of a unit of warmth the energy efficiency of the building fabric has to be taken into account (see below). It has not proved possible to break income-group expenditure down into the various uses listed in Table 8.2. Thus only heating deprivation can be examined in detail. However, it is clear that the poor are not participating in useful energy gains to the extent that other households are, either at the family or the individual level, so that on these measures, the relative state of fuel poverty is worse.

Standing charges

One of the main reasons that low-income consumers have to pay more per unit of useful energy is the quarterly charges for gas and electricity.

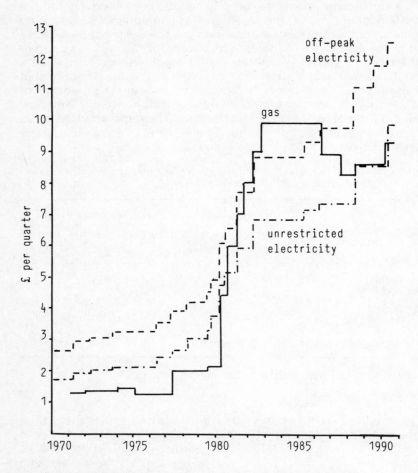

Figure 8.4 Quarterly standing charges for credit meters, 1970-90

Notes: Year date indicates January of that year.
Prior to April 1977, other gas tariffs were available for large consumers, for instance silver and gold star. Gas prices are the General Zone, the most expensive of the three covering Great Britain; at the end of 1987 the range was from £7.50-£8.30 per quarter. Electricity prices are an average of those applying in England and Wales.

Sources: Electricity Council (pers comm); British Gas Corporation (pers comm)

However truly cost reflective the standing charge may be, inevitably its impact is regressive for small users, as each extra unit consumed lowers the average cost paid, and low-income households purchase less gas and electricity than other families (Figure 8.3). This fixed charge covers the expense of getting the fuel to the customer, billing and metering, and some of the supplier's office overheads (Plumpton 1984, p.5), whereas the unit rate mainly reflects production costs.

Gas and electricity can be obtained either with a credit or a prepayment meter. There is one tariff for gas and two for electricity: unrestricted (or general tariff) electricity and a combined tariff for both on and off-peak electricity (also known as the restricted tariff). In 1970 and 1990, the relationship between the three credit standing charges was the same, with gas the cheapest and off-peak the most expensive (Figure 8.4). For a period from October 1980 there were five consecutive increases in the gas standing charge, taking the quarterly cost from £4.40 to £9.90 in two years. This rapid rise coincided with the government's required price escalation, shown clearly in Figure 2.2. However, over the twenty years, the increases are in line with inflation.

In addition to annual changes, the quarterly charges vary at any one point in time. For instance the standing charge for electricity credit meters in 1989 in Northern Ireland was 26 per cent higher than for Midlands Electricity Board customers: £13.05 and £10.34 respectively (Electricity Council 1989b, p.3). There is a similar variation between gas regions.

The regressive effect of the standing charge depends partly on the relationship between it and the unit price of fuel. The standing charges for both electricity and gas have, over the whole period, roughly reflected the rate of inflation, whereas electricity tariff charges have increased faster than the RPI and gas has become cheaper in real terms. The net effect is for intra-fuel comparison to have moved in opposite directions for electricity and for gas (Table 8.6). Over the period, the weighting of the electricity standing charge has nearly halved as it is equivalent to 0.53 GJ of electricity in 1988, instead of 0.83 GJ in 1971. For low-income consumers this trend is helpful, as a smaller proportion of expenditure goes on the fixed costs and more on useful energy. Unfortunately, the reverse happened with gas prices, and, relative to unit costs, the standing charge has risen substantially, and is equivalent to twice as much fuel as

Table 8.6 Standing charges for gas and general tariff electricity, 1971-90

	January 1971		May 1990	
	Electricity	Gas	Electricity	Gas
Cost (£ per quarter)	1.75	1.30	9.81	9.40
As a % of electricity	100	74	100	96
GJ of own fuel purchased	0.83	1.03	0.53	2.32

Note: see Figure 8.4

Sources: Based on data supplied by the Electricity Council (for England and Wales) and British Gas Corporation (for Great Britain)

in 1971. In 1988, low-income gas users spent the first 16 per cent of their expenditure on gas on the standing charge, in comparison with 11 per cent for better-off gas consumers.

The tariff rate largely reflects fuel costs whereas the standing charge is more related to fixed costs and thus returns on capital invested and labour charges. It is understandable, therefore, that there will be more similarity between the fuel utilities on the levels of standing charge increases than in tariff charges. However, the increasing proportion of income spent on gas standing charges warrants analysis. If some of the expenses at present covered by the standing charge could be reallocated to unit costs, this would reduce the regressive nature of the standing charge for low-income households. This is known as 'tariff tilting' and has been much discussed, but generally analysed only for electricity consumers (e.g. Bradshaw and Hutton 1983). Reports on the apportionment of costs to the standing charge by both Price Waterhouse and Deloitte Haskins and Sells have concluded that there is scope for a transfer of costs (Michael Fallon MP, Hansard 25 July 1983, col 840).

The full abolition of standing charges has been proposed by Members of Parliament on several occasions (ibid, col 838-53). For the poor, the general effect is assumed to be of relatively minor benefit to large numbers of small consumers in comparison with considerably greater hardship to a few large users, who are affected more by the increase in unit costs than the reduction in standing charge. Alterations to the standing charge would, therefore, make a relatively small impact on the problem of fuel poverty, but are one of the few ways in which the cost of fuel can be reduced for the poor. Since the privatization of the gas industry and the regulation of prices through formulae, there appears to be even less scope for adjustment of this kind by government or the Regulators. However, the standing charge reductions by British Gas in 1986 and 1987, whether as a result of increased efficiency or to increase competitiveness, are a welcome step in the right direction.

The standing charge is particularly disliked by pensioners and in recognition of this concern the government in 1983 required the supply industries to ensure that the standing charge represented no more than 50 per cent of the bill. The cost of this concession was estimated at £16m for electricity and £20m for gas, a year (Hansard 28 July 1983, WA col 755). This policy has now ceased, because it was found that too much help was going to the owners of second or empty homes. According to Gibson and Price there was 'no economic justification for the rebate scheme' (1986, p.262):

> It would surely be better to raise the incomes of the poor ... than to cross subsidize all users of small amounts of fuel by those who use large quantities'. (ibid, p264)

The cost of adequate additional income support (considered in Chapter 9) is extremely expensive and rarely the answer. What this proposal by Gibson and Price highlights is the continuing uncertainty about the role of the fuel industries in providing an appropriate service for all their customers, no matter what their income. This is part of the unresolved dilemma as to the responsibilities of a commercial organization that is the monopoly supplier of a basic necessity, a conflict that has not been resolved at all by privatization. Any group of customers will have special needs — the height of the meter for the disabled or the availability of service calls by appointment for working consumers. The poor would like payment methods that assist budgeting and ways to avoid the regressive impact of the standing charge. The latter are not a welfare responsibility, merely the need for the gas and electricity industries to be flexible in their treatment of all customers. This was effectively acknowledged by the Secretary of State for Energy as he compelled the supply industries to provide assistance in this instance. Thus, whilst standing charges have been an issue for some time, no satisfactory and universally acceptable solution has been found.

A slightly different aspect of the standing charge relates to prepayment meters. The cost of collecting coins from a meter is higher than the cost of reading a credit meter, so that the quarterly charge assessed on a prepayment meter is generally higher (Green et al. 1987, p.42). Coin-operated prepayment meters are being replaced gradually by token or card operated ones — they constituted 10 per cent of all British Gas's prepayment meters by March 1990. There are no coin collection costs for the utility with token meters, but there may be additional, public transport costs incurred by consumers purchasing the tokens or cards, for instance at a gas or electricity showroom or post office. The industries have not justified these higher standing charges on token-operated prepayment meters, particularly as their own costs have been lowered. The additional externalities result in a further penalty for poor users.

Even amongst low-income households, only a minority have prepayment meters: 17 per cent for gas and 11 per cent for electricity, though in both cases this represented about half of all prepayment meters (Hutton and Hardman 1990, Table 9). Prepayment meters are increasingly concentrated amongst low-income consumers, particularly those with a history of fuel debts, whereas credit meter users are found in all income groups. The rate of consumption by gas consumers with these two types of meter has diverged (Table 8.7): credit meter customers have increased consumption over the period by a greater amount than prepayment meter users. (Comparable data for electricity are not available.) This provides another piece of evidence that the poor are increasingly disadvantaged in their use of fuel, relative to other families.

Greenhouse contribution

The contribution to carbon dioxide pollution made by the different fuels results from a combination of the amount of each fuel used and its emission level. Gas is both the least polluting fuel (Table 5.3) and the greatest source of domestic delivered energy (over 59 per cent — Table 8.2). The result is that only 35 per cent of national domestic carbon dioxide

Table 8.7 Weather-corrected consumption of gas by prepayment and credit customers, GB 1970/71-1987/88 (GJ pa)

	Prepayment	Credit
1970/71	15.2	39.0
1987/88	30.6	65.4
Increase	15.4	26.4

Source: DUKES 1981, p.92 and 1989, p.70

emissions come from gas, whereas electricity contributes 45 per cent, solid fuel 15 per cent and oil 5 per cent (Boardman 1990, p.5). The most effective way to reduce pollution from carbon dioxide emissions is to reduce the 18 per cent of delivered energy that comes from electricity.

Low-income households have both a low level of total energy consumption and a greater dependence (as a proportion of income) on electricity than better-off families. The net effect is that the poorest 30 per cent of households are responsible for 24 per cent of all carbon dioxide emissions (Table 8.8). They are not the main polluters, though they offer some of the best opportunities for intervention as, for instance, all electric homes are predominantly occupied by poorer families (Chapter 5).

Table 8.8 Carbon dioxide emissions, domestic sector, by fuel and by income, UK 1987 (tonnes/household pa)

	30% of households with lowest incomes	70% other	Average
Gas	1.9	3.3	2.9
Electricity	3.0	3.9	3.6
Solid fuel	1.2	1.1	1.2
Oil	0.3	0.5	0.4
Total	6.4	8.8	8.1
% of all domestic CO_2	24	76	100

Source: Based on Boardman 1990, p.5

Expenditure by consuming households

All the evidence provided in this chapter so far has been for all fuel consumption spread across all households to give average figures. However, most households only use a selection of fuels. The poor are more dependent upon electricity and solid fuel than other households and a smaller proportion of their expenditure goes on gas, the cheapest fuel (Table 8.9). Because of this and the regressive effect of standing charges, the poorest 30 per cent consume just under 24 per cent of all domestic energy: 119,000 GWh, out of a total of 505,500 GWh (based on DUKES 1989, p.23).

Table 8.9 Fuel expenditure proportions by income, UK 1988 (%)

	30% of households with lowest incomes	70% other	Average
Electricity	50	46	47
Gas	33	42	40
Coal and coke	11	7	8
Other	6	5	5

Source: Based on FES 1988

The poor are becoming more dependent upon electricity: Table 8.10 shows the proportions of income spent on fuel in 1970. Twenty years ago the poor spent less as a proportion of income on electricity than other households and solid fuel was their main fuel expenditure. Since then, gas usage has grown, particularly in better-off homes, and electricity is now the main fuel expenditure for the poor.

Table 8.10 Fuel expenditure proportions by income, UK 1970 (%)

	30% of households with lowest incomes	70% other	Average
Electricity	36	41	40
Gas	21	25	24
Coal and coke	38	28	31
Other	5	5	5

Source: Based on FES 1970

Through a more detailed analysis of the FES it is possible to consider the trends solely amongst those households that are consumers of a particular fuel, for instance, just gas users. Expenditure by consuming households — as opposed to the same expenditure averaged over all households — gives a better indication of what is happening to individual families over time.

This analysis also highlights the cost advantage of using gas and the disadvantage of using electricity. Gas is the most important delivered energy source, as it provides 59 per cent of all domestic fuel consumption (Table 8.2). However, it represents only 40 per cent of average household expenditure (Table 8.9). Conversely, electricity provides 18 per cent of domestic fuel, but consumed 47 per cent of household expenditure. It is, therefore, extremely important, when talking about household energy use, to make it clear whether the measurement is in physical energy units or expenditure.

On electricity

In 1986, 19 per cent of the households in the 5 lowest income deciles were all electric, whereas only 10 per cent of those in the 5 highest income deciles. Hence, the poor are nearly twice as likely to be all electric than better-off families (Hutton and Hardman 1990, p.9). The distribution of electric CH by income group for the years 1976 and 1986 is given in Figure 5.4. The overall trend is for decreasing ownership: 9.4 per cent of households instead of 13.3 per cent, although the decline has been greatest amongst the higher income groups. Amongst all households with electric CH, the richest 70 per cent in 1976 were spending 43 per cent more on electricity than the poorest 30 per cent; by 1986 the differential had nearly doubled to 87 per cent (Figure 8.5). When deflated by the electricity price index, the poor are buying less electricity in 1986 than they were in 1976, whereas the rich are purchasing more.

By 1989, the average household was using the same amount of electricity (both tariffs) as in 1970 (Electricity Council 1989a, p.61), despite there being a smaller number of people in each household. As

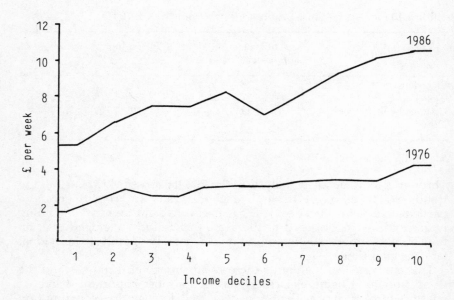

Figure 8.5 Expenditure on electricity by households with electric CH, by income, UK 1976-86

Sources: DEn 1978a, p.9; Hutton and Hardman 1990, table 3.1

demonstrated by Figure 8.3, this reflects a slight decrease in weather-corrected electricity consumption by richer families and a 34 per cent increase by low-income households. The poor still consume 23 per cent less electricity than other families, even after the decline in electric CH ownership amongst the better off. The reason for this rise in electricity use in poorer homes, but not richer ones, is not clear and could be a combination of:

— Increasing appliance ownership, particularly of those appliances that use a large amount of electricity. However, between 1970-88 all income groups increased ownership of appliances, particularly those which are heavy users of electricity, with lower income groups more likely to have obtained washing machines and higher income groups increasingly owning tumble driers and dishwashers (GHS). It is unlikely that differential changes in the ownership of electrical equipment could be the cause of an increase in consumption for low-income households, but not other families.

— Greater improvements in the energy efficiency of appliances used by better-off households, because of more rapid replacement, than amongst those in poor homes. Whilst this might contribute to the changes in electricity consumption, the low-income families, with their newer, first-time acquisitions should have fairly efficient appliances.

— Increased dependence upon electric IRH amongst low-income families. The number of households using electric IRH as the main form of heating is not known before 1976, when it was 13 per cent across all income groups though the proportion had fallen to 5 per cent in 1986 (Table 5.4). It is possible, but improbable, that this decline represents an increased concentration amongst low-income households sufficient to explain the increased electricity consumption.

— The greater decrease in the size of better-off households (from 3.40 to 2.96 persons per household) in comparison with poorer ones (from 1.90 to 1.75 persons per household).

There is, therefore, no clear explanation for the increase in electricity consumption amongst low-income households in comparison with the decline amongst other families. Because on-peak electricity is the most expensive form of useful energy, this trend is to the disadvantage of the poor.

On gas

The number of households connected to the gas supply has risen from 68 per cent in 1970/1 to 78 per cent in 1990 (British Gas Annual Reports), with the biggest increase occurring between 1978-80, the end of the natural gas conversion programme. Between 1976 and 1986 (when the total connected was 74 per cent), the majority of new connections were to households in the top 60 per cent of the income range (Figure 5.2). The amount spent by households with gas CH, by income, in 1976 and 1986 (the greatest span of years available) is shown in Figure 8.6. In both years, the differential across income groups was the same, with richer families spending a third more than the poor. When expenditure is deflated by the gas price index, both groups have slightly increased their real gas purchases. In absolute terms, the gap between the two income groups has widened, as 30 per cent increase of a large number is obviously more than the same proportional addition to a smaller number. Thus, the amount of gas being consumed by better-off families has increased by more over this ten-year period than consumption in poor families.

The data from Chapter 5 and this chapter combine to show that low-income families, in comparison with other households, are less likely to be connected to the gas supply, and if connected they are less likely to

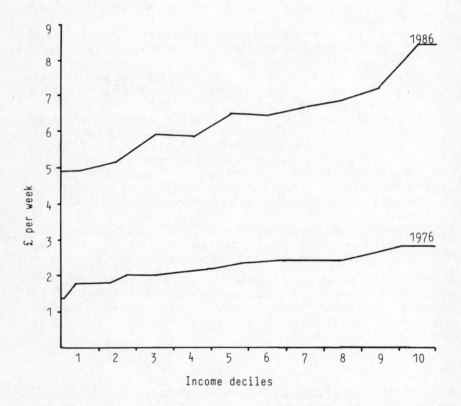

Figure 8.6 Expenditure on gas by households with gas CH, by income, UK 1976-86

Sources: DEn 1978a, p.9; Hutton and Hardman 1990, Table 3.1

have gas CH, and if they have gas CH, they purchase considerably less energy, both delivered and useful than better-off families. All three of these trends are static or strengthening; the lot of the poor, in relation to other households, is not improving. The poor have difficulty affording to use even the most energy efficient heating system.

Trends in solid fuel consumption and expenditure are unobtainable, because of the poor coverage of these payments in the FES (see FES and Fuel Costs in Glossary). In 1986, 13 per cent of pensioner households used solid fuel in comparison with 9 per cent of non-pensioner households (Hutton and Hardman 1990, p14). 'Other fuels' include paraffin, liquid petroleum gas, wood and oil for CH, which, in different combinations, are of similar importance to most income groups and are used by an average of 8 per cent of households in 1986 (ibid, p.8). However, data are too sparse to provide identifiable trends.

In summary, low-income households have been increasing their purchases of useful energy at a slower rate than other families and reducing the cost per gigajoule more slowly. This can be explained partly by a slower rate of fuel substitution in poor households, with gas being used to replace all other fuels, and partly by an increase in the use of general tariff electricity, the most expensive fuel. These two factors are closely interdependent, particularly in low-income homes: increased consumption is constrained by the inability to reduce unit costs.

The trends with total useful energy are mirrored in the use of specific fuels for heating, as the differentials are widening between low-income and other consumers with both gas and electric CH. For both gas and electricity users, the relationship between expenditure and consumption is distorted by the existence of standing charges.

Useful energy and warmth

In 1988, poor households, in comparison with other families:

— required 11 per cent less room-hours of warmth (Table 6.4), if the dwellings they occupied were of the size assumed in the heating standard. If the match was less accurate, as seems likely, the number of room-hours of warmth was probably more nearly the same across the income groups;
— bought 32 per cent less useful energy (Table 8.5).

Low-income households contain fewer people than the average, so that a lower level of consumption might be expected, but the evidence on heating needs indicates that energy use is a combination of a substantial initial level of service, together with smaller increments for each additional member of the family, probably applying to other forms of energy service as well. As there are no national data on building fabric heat loss by income, NCH homes have been taken as a proxy for low-income households:

— in 1988, NCH homes had a heat loss rate 1 per cent higher than that in CH homes (Table 4.6).

These three pieces of evidence combine to show that low-income households in NCH homes are colder than higher income families in CH homes in 1988: the level of demand in NCH is 10 per cent lower (89 per cent x 101 per cent), whereas the variation in consumption is 32 per cent. Therefore, in 1988, the poor were able to purchase 22 per cent (32-10 per cent) less warmth than they needed in comparison with other households. Between:

— 1970-88, purchases of useful energy increased by 20 per cent in low-income households and by 28 per cent in other households (Table 8.5);
— 1976-88, the rate of heat loss in the average house dropped by 17 per cent in NCH and 21 per cent in CH ones (Table 4.6). Prior to 1976, most loft insulation had been installed by owner occupiers.

For these reasons, between 1970-88 the level of warmth in better-off households increased more than in poor homes, so that, relatively, the poor are colder in 1988, than they were in 1970. Whilst it is a simplification to equate low incomes with NCH homes and high incomes with CH homes, there is sufficient overlap to justify the assumption (Figure 5.1). Also, the data are not available to give more consistent definitions. Therefore, the suspicion raised in the previous chapter of an increasing divide between the temperatures in CH and NCH homes appears to be confirmed. Fuel poverty is getting worse because the poor are colder than other families and increasingly so.

These conclusions can be made despite several assumptions which probably penalize the poor:

— the same technical efficiencies have been applied to consumption by both income groups, though in practice, the poor are likely to have less efficient appliances, because of age or disrepair;
— the same cost of fuel has been used across the income groups, although the poor often have to pay more for the same fuel, for instance because they use prepayment meters;
— the model of heat losses in the housing stock makes no allowance for the effect of disrepair on the building fabric heat losses.

When less aggregated data are used, as for instance on the amount of gas consumed by households with gas CH, there is evidence of a decrease in the quantity purchased by low-income households over time. If this is not matched by purchases of other fuels, fuel poverty is worsening in absolute and relative terms for some households.

As total heat loss in both CH and NCH homes has declined by approximately similar amounts (Table 4.6), the reason for the greater increase in warmth in most homes is the rise in useful energy. As noted earlier, there have been government-funded schemes to improve the energy efficiency of low-income homes through insulation measures, resulting in expenditure over 1978-89 of just over £300m (Table 4.5). Nearly a third of all the money spent on low-income homes has been through the local authorities energy conservation programme. There has been no comparable assistance with fuel substitution. If low-income households are warmer in 1990, it is largely because central and local government capital expenditure programmes have enabled them to approach better-off households in installation of basic thermal insulation improvements. In the absence of assistance with improvements to the energy efficiency of heating systems, the poor have lagged behind other households and increased their purchases of useful energy by a smaller amount.

Precise cross-checking of these findings on useful energy, rates of heat loss and warmth levels requires better data and a detailed and complex computer program, similar to the one developed by Leach and Pellew (1982). An exercise to estimate the warmth deprivation in low-income households was undertaken by Aird using 1975 expenditure data. He established that each unit of warmth cost the poorest 11 per cent of households 63p per therm equivalent, in comparison with 46p for average families and 40p for the richest 13 per cent: a 'consumer fuel detriment' of 36 per cent for the poor and a bonus of 14 per cent for the rich (Aird 1977, pp11-12). This is considerably greater than that estimated above but as the poor have been defined as different proportions of the population by Aird and in this book, no comparison can be made between the two findings, except that each reinforces the other.

The temperature improvements calculated to have occurred in CH and NCH homes are given in Table 7.1. As stated, Henderson and Shorrock (1989) assume an equal distribution of additional useful energy purchases as well as heat loss reductions between CH and NCH homes. Hence, the standard difference of 2.5°C between their calculated temperatures. As this chapter has shown that useful energy purchases have increased disproportionately between the rich and the poor, there must be some doubt about the calculated temperature increases. It seems almost certain that temperatures in CH homes have risen faster than those in NCH. Average increases in useful energy purchases appear to mask a considerable variation between the greater increase in warmth in better-off CH households and the smaller or nil improvement in temperatures in the NCH homes of poor families.

The conclusion reached by Abel-Smith and Townsend about poverty (as quoted in Chapter 2) appears to be appropriate for heating as well, though it cannot be put so forcibly:

> the data we have presented contradicts the commonly held view that a trend towards greater equality has accompanied the trend towards greater affluence. (1965, p.66)

Conclusions

Low-income households buy less useful energy and pay more per unit than other families and these inequalities have worsened since 1970. The main reason for the increasing disparity is the slower rate of fuel substitution over time by the poor. That is the dynamic effect. The static effect is the lower absolute expenditure by the poor in any one year together with the regressive impact of standing charges at low levels of consumption. The poor are buying more expensive useful energy and, therefore, more expensive warmth.

The evidence on useful energy purchases confirms the tentative finding in Chapter 7 on temperature trends, that improvements are occurring unequally, with increasing disparity between CH and NCH homes. The gap between low-income and other households may or may not be as great, but it is certainly widening and meant that in 1988 poor families were able to purchase 22 per cent less warmth than they needed in comparison with other households. A similar comparison undertaken with 1985 data showed that the gap in that year was only 18 per cent, indicating a divergent trend. Thus, the evidence from Chapters 4-8 can be combined plausibly to show that the poor have to buy more expensive energy and have colder homes than other families and that the differences are growing. Fuel poverty is increasing. The situation would have been worse but for the ameliorating impact of government-funded insulation programmes.

In the average household and at the total national level, the energy efficiency improvements that have enabled an increase in the purchases of useful energy have also resulted in lower consumption of delivered energy. When the cost of warmth is lowered through energy efficiency improvements, the temperature inside the house increases although purchases of delivered energy drop. For low-income households to obtain similar levels of warmth as other families, or at least the level of the heating standard, there needs to be a more positive bias in government policy towards the poor, to include assistance with fuel substitution. Then the growth in fuel poverty can be reversed. At the moment, the poor are constrained in their ability to purchase more useful energy by their inability to reduce the unit cost of useful energy, particularly through fuel substitution. Both increased energy efficiency and fuel substitution are necessary components of policies to reduce domestic carbon dioxide emissions, particularly focusing on electricity use.

9 The value of warmth

The value placed on warmth is important at both the household level and in national decision-making, though it appears to have been little discussed. Families that suffer from fuel poverty can be expected to spend more on warmth if it becomes cheaper, or if their income increases. Thus, the extent of fuel poverty can be confirmed from the strength of the evidence of additional demand for warmth amongst low-income households. However, this latent demand for extra warmth undermines any policy where the criterion for success is measured in fuel savings (i.e. energy conservation). There is conflicting evidence from government departments both about whether the value of additional warmth should be included as a benefit and, if so, how that value should be calculated in practice.

In the area of domestic energy, the emphasis on market forces is meant to ensure that the optimal balance is achieved between energy supply and demand and, for fuel poverty, there are additional judgements about revenue support instead of or as well as capital expenditure. The debate on costs and benefits at this level is, therefore, not just a semantic discussion. At issue is the proper allocation of resources totalling billions of pounds, whether for energy supply through the privatized supply companies or DEn, energy efficiency improvements through the DOE or income support by the DSS. As a basis for policy and investment decisions, the findings of this chapter are of considerable importance and provide a stepping-stone for the proposals in Chapter 10.

Economic models that attempt to distil and explain the reality of household behaviour — positive economics — provide the basis for assessing policy options. The factors that influence household demand for warmth and the value placed on it can be assessed through economic theory. With the assistance of government-sponsored research into the effect of energy efficiency improvements, one model is developed into a predictive tool to establish a procedure for putting a monetary value on warmth.

As explained in Chapter 1, the Cost of Warmth Index provides a structure to relate the seven main factors to one another. As a simplified form of energy audit, it enables technical, social and economic parameters to be combined for an individual household. The result is essentially a summary of a static situation — the cost of keeping warm at a point in time. In the last chapter, expenditure data are used to indicate how changes in the cost of useful energy have affected the demand for energy, over time. This introduces a dynamic element into the assessment of fuel poverty. The further development of this evidence, through economic models, enables the effect of policy options to be analysed and predicted with greater accuracy. The other real benefit to accrue from this analysis is that it measures people's own assessment of their need for additional warmth — the value they place on it, rather than the important, but abstract, concept of thermal comfort discussed in Chapter 6.

Most of the description concerns the value placed on warmth. The discussion is just as relevant to other uses of energy in the home, but these are less complicated because they only involve one set of capital goods, not two.

Elasticities

As described in Chapter 8, increased consumption of useful energy has occurred in both low-income and other households over the period 1970-88. In terms of household economics, three different factors could have encouraged this additional consumption:

— a reduction in the price of useful energy, either through fuel substitution or a fall in the real price, as has occurred with gas;
— a reduction in the price of warmth through insulation measures;
— an increase in real incomes.

With the data presented so far, it has not been possible to differentiate between these three effects. For fuel poverty policy purposes, it is important to be able to distinguish between the impact on consumption of reducing the cost of warmth and the effect of an increase in income, and the way these interact.

Economists distinguish between the effect of changes in price and changes in income through elasticities: an elasticity measures the degree of responsiveness of one variable to changes in another, quantified as proportions of the original values. The demand for a commodity can be assessed in relation to changes in the price (price elasticity of demand) or income (income elasticity of demand). The values of income elasticities of demand, have the following implications (Lipsey 1979, p.108):

0	perfectly or completely inelastic	quantity demanded does not change as income changes
$>0<1$	inelastic	quantity demanded changes by a smaller percentage than does income
1	unit elastic	quantity demanded changes by exactly the same percentage as does income
$>1<\infty$	elastic	quantity demanded changes by a larger percentage than does income

The general principle with a basic necessity, like warmth, is that low-income households (if there is substantial deprivation) will have a higher income elasticity than other families, who are already purchasing more warmth. The demand for warmth is limited to comfort conditions, beyond which there is no further benefit, so that the income elasticity is close to zero in an adequately warm household. Thus, the income elasticity is high if there is deprivation and gradually reduces to reflect increasing satisfaction with the level of warmth being achieved in the home at present. Warmth is unusual in this respect — with food or eating out, for example, additional expenditure continues to give further utility.

Hutton *et al.* (1987) have undertaken research using FES data and subdivided households into three income groups according to the income elasticity they exhibit for fuel expenditure. For all households, there is a band of about 15 per cent of households in the middle of the income range (37-52 per cent) with an income elasticity close to zero for fuel expenditure. Households with lower incomes (i.e. the bottom 37 per cent) have an income elasticity of 0.59 and those with a higher income (the top 48 per cent) have an income elasticity of 0.40 for fuel expenditure. It is not clear why higher income families should want additional energy purchases more than middle income households, but the highest income elasticity is found in the poorest group, indicating that additional fuel expenditure has a relatively high utility for them. Table 9.1 gives comparable information for some of the groups within the population

Table 9.1 Income elasticities for fuel expenditure by income and household characteristics for lower income band*

	Proportion of whole group %	Income elasticity for fuel expenditure
Single pensioner	81	0.96
Two pensioners	31	0.76
Single parent	51	0.98
Two adults + 1 child	18	0.74
Two adults + 2 children	16	0.72
Two adults, 3+ children	27	0.66
Whole sample	37	0.59

Note:
* There is an 'unconstrained' group of households approximately in the middle of the income distribution who have an income elasticity for fuel of about zero. The information in this table is for households with incomes lower than those in the unconstrained group. Not all groups are listed, for instance adults without children, which is why the elasticity for the whole sample is lower than that of the groups listed.

Source: Hutton *et al.* 1987, p.13, using 1982 FES

showing, for instance, that the 81 per cent of single pensioner households with the lowest incomes had an income elasticity of demand for further fuel expenditure of 0.96 in 1982. Single pensioners currently spend 13.4 per cent of their weekly expenditure on fuel (Bradshaw *et al.* 1986, p.36). This means that if the national insurance pension for a single person were increased by 10 per cent (£4.00 per week in 1988), this would result in pensioners spending an extra 51p a week expenditure on fuel, to bring the total up to £5.74. Pensioners and single-parent families have the highest income elasticities, showing that these two groups are more constrained by their present incomes than other households and have the greatest need for additional warmth.

Hutton's evidence appears to conflict with the Dilnot and Helm (1987) finding that there is almost no growth in fuel expenditure with the rise in income when standardized for family composition. Although they draw their evidence from the 1982 and 1984 FES respectively this should not be the reason for the discrepancy. Part of the explanation is that the nine income bands used by Dilnot and Helm are unrelated to proportions of the population. With pensioner couples for example, it appears that the top seven bands represent less than 30 per cent of this group of households: the majority of the evidence concerns a minority of the population and these are the richest households. As explained earlier, Hutton found that rich households have higher income elasticities than households in the middle of the income range, which is both counter-intuitive and in further conflict with Dilnot and Helm. However, on balance, for low-income families, the Dilnot and Helm statement does not appear to invalidate Hutton's findings, which can still be used. Without further analysis, these ambiguities can only be acknowledged and ignored.

The average of 0.59 found by Hutton *et al.* is considerably lower than than the DEn's estimate that the income elasticity of demand for domestic energy (still excluding transport energy) is 0.9 (1983b, Appendix III). The DEn are implying that the average British household has a high demand for extra energy.

The 30 per cent of households with the lowest incomes spent £8.48 on fuel per week in 1988, 10 per cent of their total expenditure (Table 3.4); expenditure by the remaining 70 per cent of households was £11.34. Because of the problems in obtaining comparable income data for these two groups, it is assumed that total expenditure is equivalent to net income, so that income elasticities can be applied to extra expenditure. This introduces less of a distortion amongst low-income households, where savings are minimal. If differences in energy efficiency are ignored, to raise fuel expenditure in poor households to that of other families (an additional £2.90 per week), with an income elasticity of 0.59 (from Table 9.1) would require total expenditure in low-income households to increase from £84.56 per week to £129.24 — a rise of 53 per cent. To give the 30 per cent of households that are dependent upon the state an additional £44.68 per week (1988£) would have increased the social security budget by over £15bn in 1988. This is equivalent to an extra

9 per cent on present total public expenditure of £167bn and is required indefinitely. As this figure ignores the higher cost of warmth in low-income homes, it underestimates the public expenditure necessary to provide the poor with sufficient income to keep warm. Even with high income elasticities (relative to the rest of the population), this demonstrates that additional income support would be an extremely expensive way of combating fuel poverty and politically an unrealistic solution.

The use of income elasticities together with the proportion of expenditure on fuel reinforces evidence on fuel poverty. In Chapter 3, the Isherwood and Hancock study (1979) found that 18 per cent of households had disproportionately high expenditure on fuel. They established that households spending less than twice the median on fuel (as a proportion of income) had lower income elasticities for electricity (0.33) and gas (0.56) than those that already spend a disproportionate amount on fuel, with elasticities of 0.57 for electricity and 0.73 for gas (ibid, p.27). The implication is that those who are spending a relatively large amount of their income on fuel are still cold: they are not spending a lot in order to make sure they are warm, as purchasing extra warmth is still a high priority. Thus, households spending a large proportion of their income on fuel still have cold homes, demonstrating that fuel poverty does not manifest as either high expenditure or low temperatures — it is commonly experienced as both.

The interpretation of price elasticities of demand is similar to that of income elasticities, but more complex. A fall in the price of a commodity means that the same quantity is cheaper to buy, so that the household has extra money to spend, either on more of the same commodity or on alternatives. There are, therefore, two effects combined in a price elasticity of demand: the substitution effect of the lower price of the commodity and the income effect of the additional income that this gives the household. The income effect is of the same magnitude as the income elasticity for the same household.

The commodity to be purchased, in this context, can be fuel (that is delivered energy), useful energy or warmth. In the COWI, the price of fuel is one of the seven factors (COWI 4), the cost of useful energy comes from two factors (COWI 4 ÷ COWI 3) and the cost of warmth reflects the first four factors (Table 1.1). Therefore, for fuel poverty, the price elasticity of demand for delivered energy, useful energy and warmth are of increasing significance because they successively include more variables.

Because of the additional complexity of price elasticities and the interdependence of several factors in determining the demand for warmth, the evidence on the price elasticities for warmth amongst low-income households is much more circumstantial than that for income elasticities. Some evidence can be derived from research undertaken by the BRE in Scottish local authority housing relating delivered energy consumption to the rate of heat loss in the dwelling (Figure 9.1). Although no data are given on income levels, probably at least half of the households in the study were poor — the average for this group across the UK. For the average energy users in this study, about half the potential benefits of improved insulation were taken as energy savings, whereas the top 10 per cent of users took three-quarters of the benefit as savings (Cornish 1977, p.3). These figures indicate price elasticities of demand for warmth of -0.50 and -0.25 respectively. For the whole domestic sector, the DEn estimate that the:

> short run price elasticity of demand for space and water heating is about -0.2. This suggests that up to 20% of the reduction in heating bills could be re-spent on space heating. (DEn 1983b, Appendix III)

The higher price elasticities in Cornish's study would indicate that they are low-income households, colder than the national average.

If a line showing the consumption necessary to maintain equal warmth was added to Figure 9.1, it would be above that of the top 10 per cent of households and have a steeper gradient. Therefore, even the 10 per cent of households with the highest energy consumption do not increase purchases of delivered energy sufficiently to combat the effect of increasing heat loss and maintain equivalent warmth. With all the other households there is a greater lowering of temperature as energy inefficiency increases. All these local authority households are demonstrating a demand for extra warmth when the cost is lowered through better insulation and in many cases this demand can be satisfied despite decreasing consumption of delivered energy. This finding is the same as the general trend identified from national statistics in Chapter 8: decreasing delivered energy consumption per household has occurred, whilst useful energy purchases and temperatures have risen.

For any given level of warmth, the demand for delivered energy is inversely associated with the energy efficiency of the building fabric (nothing is known about the efficiency of the heating systems in Cornish's sample). However, as the wide spread of energy consumption at high rates of heat loss demonstrates, other factors influence household demand for delivered energy. For the 10 per cent of consumers with the lowest consumption, the higher the heat loss (and by implication the greater the cost per unit of warmth) the less energy they purchase. The low consuming families in the most poorly insulated properties, where the cost of warmth is high, are using so little energy that they must be extremely cold. For them, additional purchases of fuel in their present homes have a low utility, apparently because they obtain so little benefit from the expenditure. However, when the energy efficiency is improved through increased insulation, these families purchase more delivered energy. They are warmer both because of the added insulation and as a result of the extra energy consumption. To the extent that this measures the warmth achieved, it is evidence that the price elasticity of demand for warmth is highest amongst the coldest families.

Although no information is available on the cost of warmth or of delivered energy in the study by Cornish, an increased demand for additional warmth as a result of energy efficiency improvements can be

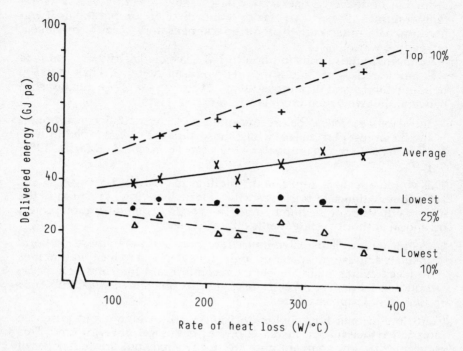

Figure 9.1 Annual energy consumption, per household, in relation to the rate of heat loss, local authority dwellings

Notes: The rate of heat loss was measured in a sample of 1,800 dwellings, many of which had been insulated. The lower the heat loss, the less energy is required to achieve a specific level of warmth. Within each group of houses with the same heat loss, the dwellings were ranked by level of energy consumption, so that each data point is the mean for at least 20 families. The rate of heat loss excludes ventilation losses and cannot, therefore, be compared with data in Chapter 4, for instance Table 4.6.

Source: Cornish 1977, p.8, BRE Crown Copyright 1990

accompanied either by a decrease (normally) or an increase in delivered energy purchases. It is important to distinguish between the price elasticity of demand for warmth and the price elasticity of demand for delivered energy, as they can be of quite different magnitudes in the same household.

An increase in the price of fuel is equivalent to an increase in the rate of heat loss — both put up the cost of warmth. However, a fuel price increase moves people down the vertical axis instead of along the horizontal one. When the price goes up, for each £ of expenditure on fuel, less delivered energy is purchased. Thus, any price increase results in a reduction in delivered energy purchases and a lowering of household temperatures. These two factors are directly proportional if no investment is undertaken to produce a compensatory lowering in the cost of keeping warm.

In another study, consumption in 100 Glasgow District Council flats was monitored before and after energy efficiency improvements to the building fabric and heating system. There was a close relationship between the two expenditure levels, but:

> those with the *lowest* 'before' expenditures *increased* their consumption by the largest percentage, whilst those with the *highest* expenditures, decreased their consumption by the largest percentage. (Markus 1987, p.9)

This would appear to replicate the findings in Figure 9.1.

Further evidence of price elasticities for fuel varying in relation to the energy efficiency of the home comes from an EEO demonstration project on a local authority estate that is difficult to let:

> People with well insulated houses are more likely than those in poorly insulated houses to spend enough money on fuel to heat their homes to levels which maintain personal comfort and the condition of the building fabric, *since they perceive such spending as effective.* (EEDS 56 1988, p.3. Emphasis added)

Cold, low-income households do value warmth highly and purchase more when their income increases or the price of warmth is lowered. This confirms that low temperatures are not desired, but are indicative of economic constraint and fuel poverty. For many of these households, the greatest increase in warmth occurs as a result of energy efficiency improvements, whereas fuel price increases result in decreased consumption.

Modelling demand

The relationship between energy consumption and consumer well-being has been modelled by Barnett (1986), incorporating the effect of different elasticities for energy services, such as warmth (Figure 9.2). Income elasticities are not involved. He demonstrates that alterations to the energy efficiency of the home always result in positive improvements to the well-being of the occupants, whether or not there has been a reduction in energy use, because his model enables the value of warmth to be included. Household expenditure is represented in quadrant 3 and this is converted into a quantity of energy service (quadrant 1) by the technical relationship between delivered energy and energy service, symbolized by the line t0t', for that household's lifestyle. Improvements to the energy efficiency of the home result in the line t0t' pivoting anti-clockwise to e0e' (a clockwise rotation would be caused, for instance, by the need to keep more rooms warm). The way the household responds to these energy efficiency improvements depends upon its elasticity of demand for warmth:

— the household was adequately warm (α) and so has zero price elasticity for additional warmth and the same quantity of energy service is purchased (ϵ). Because of the increased energy efficiency of the dwelling, the same level of energy service is obtained despite lower expenditure on delivered energy (e);
— when demand is price inelastic, less fuel is purchased (d), though there may still be an increase in energy services (δ);
— price elasticity of demand for energy services is 1, so that consumption of energy services increases to β, which can be achieved with the same expenditure on delivered energy (a);
— demand is price elastic when expenditure on the energy service increases by a higher percentage than the price falls. As a result of increased expenditure on delivered energy (c), and the increased energy efficiency of the dwelling, the energy services rises to γ.

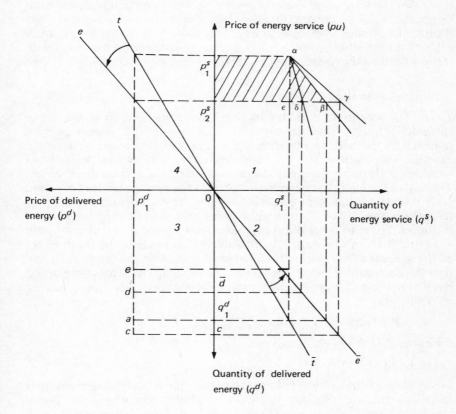

Figure 9.2 Delivered energy and consumer well-being

Source: Barnett 1986, p.428, with point e added

In each case, the new line αε, αδ, αβ or αγ is part of the demand curve for energy service of that particular household. With all these cases there is an increase in consumer surplus — the shaded area of the rectangle up to αε, plus the appropriate triangle — the benefit that results from the price reduction in the cost of a unit of warmth. In all cases except the first (ε), the maximum energy saving is offset by the cost of purchasing additional units of warmth (a vertical, unshaded rectangle) to obtain the net benefit to the consumer. Thus:

> the greater the price elasticity of demand for energy service, the greater will be the gain in household wellbeing following the introduction of an energy conservation measure (ibid, p.429),

and the larger the proportion that is taken as additional warmth. The greatest increase in consumer benefit actually occurs when expenditure also increases, as demonstrated by Cornish's work above (Figure 9.1). In this case, the additional value of warmth (vertical rectangle down from εγ to the axis) is greater in area than the value of energy savings (horizontal shaded rectangle to αε). This is why 'it is inappropriate to discuss the benefits to households of energy conservation measures in terms of reduced expenditure on delivered energy' (ibid, p.429).

Research confirmation

To test the validity of Barnett's model, the results from six of the Better Insulated House schemes have been analysed. The energy service required is defined as the increase in internal temperature over the external temperature. The free incidental gains — solar and metabolic — would slightly narrow this gap, but are not quantifiable from this data set. The cost of the energy service is measured as the average cost of achieving each of these additional degrees (£/°C), using expenditure on all energy — this allows the cost of the incidental gains from non-heating usage to be included. The results for Whitburn have been plotted for quadrant 1 only — Figure 9.3 — and Figure 9.4 and Table 9.2 summarize the findings for all the schemes. The static situations represented by the original (0) and new (N) choices are both on the occupants' demand curve for warmth and have been plotted from the survey data. Of the maximum energy saving at Whitburn:

$$(66 - 44) \times (12.9 - 6.3) = £146$$

the consumers have chosen to use:

$$(14.4 - 12.9) \times 44 = £66$$

to purchase additional warmth, so the net saving in energy expenditure is £80. This finding also comes from Campbell's Report on the Better Insulated House programme (1985, Table 1.2). The remaining portion of the consumer surplus, the triangular area (£17) together with this recycled energy expenditure represents the value of warmth to the households (£83).

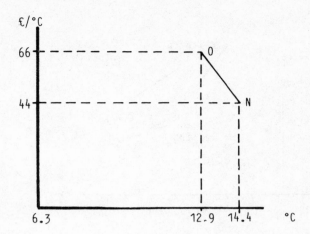

Figure 9.3 Annual benefit of energy efficiency inprovements, Whitburn, Better Insulated House (BIH) programme

Source: Based on Campbell 1985

The sections of the demand curves from each of the schemes are amalgamated (Figure 9.4), showing consistently similar trends. As the homes are made more energy efficient, the cost of each additional degree of warmth becomes progressively less, and extra warmth cheaper to obtain. Thus, improvements to homes that are already relatively energy efficient can result in significant temperature increases. At some point, probably about 21°C, an inflexion can be expected in the curve, as above this level, additional warmth is dysfunctional. None of these local authority schemes had reached that point. Conversely, in homes that are expensive to heat, little warmth is bought and the demand curve is nearly vertical close to the external temperature (plus free incidental gains). When the temperature gain is relatively small, perhaps 2°C or less, then the shape of the demand curve does not materially affect the value attributed to the extra warmth. If the demand curve is concave and a large increase in temperature occurs, this method slightly overestimates the value of the warmth. The cost of warmth in these schemes varied from £66/°C before improvements at Whitburn to £22/°C after improvements at Coventry — a threefold variation.

For the 120 households in improved houses in these six BIH projects, just over a third of the benefit of the greater energy efficiency was taken as energy savings and 61 per cent as additional warmth (Table 9.2). In none of the schemes was more than half of the benefit taken as energy savings. However, in all the projects, some energy was saved. The coldest homes, before alterations, were those where the cost of warmth was highest

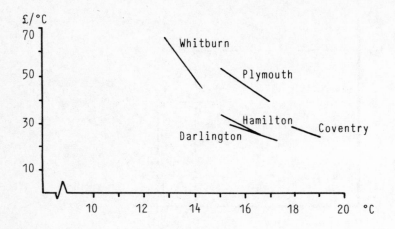

Figure 9.4 Demand curves for warmth from Better Insulated House (BIH) Programme

Source: Based on Campbell 1985

(Whitburn and Plymouth) and where the greatest temperature increase occurred after the energy efficiency improvements. In other words, the additional warmth is valued most highly by people in energy inefficient homes. The incomes of the occupants are not known, but as half of all local authority tenants are poor, the results can be taken as indicative of the effect on low-income families. The proportion of benefit taken as additional warmth might be even higher in a sample of homes occupied solely by the poor. The most energy inefficient housing is in the privately rented sector (Table 4.6) where the highest price elasticities for warmth can be expected.

In American homes, after energy efficiency improvements, 'low-income households were found to increase their temperature levels significantly more than their non-low-income counterparts' (Dinan and Trumble 1989, p.46). Economic theory could have predicted this because the income effect is greater when 'heating costs … constitute a larger share of the household budget' and the lower temperatures in low-income households mean they 'will have more motivation to choose higher temperature levels as the cost of heat is reduced' (ibid, pp.46-7).

With this suggested method of valuing warmth, the greatest benefit has occurred in homes with the largest improvement in energy efficiency: Whitburn and Plymouth. Previously, the Plymouth scheme appeared to show a disappointing result because of the low reduction in expenditure and thus it would not have been deemed successful against conventional criteria based on energy savings, precisely because the elderly occupants

Table 9.2 Assessment of the economic benefits from energy efficiency improvements to six of the Better Insulated House schemes

	Whitburn	Hamilton	Plymouth	Darlington	Coventry s g*	Coventry d g*	Mean
Savings on energy costs (£)	80	32	16	31	18	16	32
Value of warmth (£)	83	46	92	33	31	16	50
Total benefit (p.a.) (£)	163	78	108	64	49	32	82
% as savings	49	41	15	48	36	50	39
% as warmth	51	59	85	52	64	50	61
Original temps (°C)	12.9	15.1	15.0	15.4	17.8	17.8	15.7
Original cost per °C (£)	66	33	53	29	28	28	40
Temperature gain (°C)	1.5	1.6	2.0	1.3	1.2	0.6	1.4
Value per degree gain (£)	55	29	46	25	26	27	36

Note:
* s g: single glazing; d g: double glazing

Source: Based on Campbell 1985

value additional warmth highly. By including the value of warmth, good payback periods for all schemes except the double-glazed Coventry project have been obtained (Table 9.3): these are crude paybacks, not discounted cash flows (both in Glossary). Cost effectiveness would be increased if other benefits, such as reduced condensation and deterioration to property, could be included.

Having shown the need to include a value for the additional warmth and demonstrated from Barnett's model how this can be done, the

Table 9.3 Payback periods from energy efficiency improvements to six of the Better Insulated House schemes (1982£)

	Whitburn	Hamilton	Plymouth	Darlington	Coventry s g*	Coventry d g*	Mean
Capital cost (£)	216	103	267	53	83	267	165
Benefit (£)	163	78	108	64	49	32	82
Payback (years)	1.3	1.3	2.5	<1	1.7	8.3	2.0

Note:
* s g: single glazing; d g: double glazing

Source: Line 1 — Campbell 1985, col 9, Table A.1; Line 2 — Table 9.2

economic analysis can now be related to the technical methods at present being used by other researchers in the field.

Present methods

Families take the benefit of improved energy efficiency as some combination of increased temperature and reduced expenditure and the economic model provides a basis for putting a monetary value on this additional warmth. Government departments and researchers have varying attitudes to the benefit and valuation of warmth. Whilst additional warmth has been cited as a benefit (DOE 1973, p.1; ACEC 1983, p.15), the clearest statement about the importance of including a value for warmth in economic assessments has been made by a committee set up by the government, under the auspices of the DOE. The following quotation is taken from their third publication, known as DEN 3 (Domestic Energy Notes 3) — a report referred to in public sector housing design guides (e.g. Housing Corporation 1985):

> where the landlord is a public body all the benefits should be counted, including those to the tenant, because what matters in the expenditure of public funds is simply whether the total sum of benefits is greater than that of costs, whoever receives them... Experience suggests that a large proportion (perhaps a half) of the potential fuel savings from better insulation standards, or more effective heating systems, are not in fact made because occupiers decide to keep their dwellings warmer. This implies that the benefit to them of the warmer dwelling is at least as great as that of the even lower fuel bill they could otherwise have, because they are free to choose between the two. It is therefore reasonable as a broad approximation for the purposes of the calculation, to assume that all the benefit accrues as fuel savings, with the proviso that this will probably underestimate to some degree the total benefits to the tenants, but substantially overestimate the actual energy saving. (DEN 3, 1978, p.14)

This excellent statement makes it quite clear that the well-established principles of cost benefit analysis should be followed and the value of warmth included in cost-effectiveness calculations of energy efficiency improvements. However, there is still some ambiguity in the statement about how to put a monetary value on that additional warmth. As a result, there are several different methods in use at present involving different permutations of the main variables: the energy efficiency (usually rate of heat loss), internal temperature, the amount of energy used and energy expenditure of the old and the new conditions. Some of the scenarios that have been used in different reports are built up from the components in Table 9.4. None of these include the cost of the fuel, either before or after the improvements, even though it is of crucial importance. This is partly because of the original emphasis on energy conservation in government, so that most assessments have used physical energy, not monetary units. This multiplicity of present methods demonstrates why

Table 9.4 Components of different scenarios for measuring warmth

	A Original	B Final	C Final building, original temperature	D Original building, final temperature	E Final building, original energy
Heat loss rate (W/°C)	400	300	300	400	300
Temperature gain (°C)*	6.0	7.5	6.0	7.5	8.0
Energy used for heating (W)**	1200	1050	600	1800	1200

Notes:
* Gain over external temperature; ** Uniform incidental gains of 1200 W are present, thus reducing the total energy needed for heating to the level given in the Table.

Source: Campbell 29 Aug. 1986, pers comm

there is a need for a consistent approach, as defined by the economic model already described, utilizing Barnett's work.

A vs. B — a simple comparison showing small fuel savings and noting, but not evaluating, the additional warmth;

B vs. D — the appropriate comparison where there is no change in the demand temperature, because the family were adequately warm beforehand. This method is used in energy auditing programs, such as BREDEM and Energy Auditor, with the addition of the cost of the fuel for cost-effectiveness calculations.

A vs. C — BRECSU (for the DEn) estimate the amount of energy that would have been saved if the test houses, after improvements, had been heated only to the temperature previously achieved in the control houses. This, in effect, ignores the temperature gain (EEDS 1983, p.15; EEDS 1984, p.11);

B vs. E — Birmingham's Housing Department and the DOE combined on the EIK project and obtained the proportion of benefit taken as warmth by calculating what the temperature in the control houses would have been if no energy savings had occurred, and expressed this in energy units (EIK Project 1982, Table 3.5b);

A vs. B and B vs. C — the DOE's Better Insulated House programme assesses energy savings and greater warmth separately and in energy units. Energy savings, from the straightforward comparison of original and final (A vs. B), are used in calculating payback periods. The energy equivalent of additional warmth is obtained by comparing B with C, but is not used as part of the total benefit (Campbell 1985). Had these been amalgamated, the result would be A vs. C, the same as BRECSU;

DEN 3 — Sheldrick's is the only attempt at giving warmth a monetary value, by following the recommendation in DEN 3 literally. Having calculated the value of the reduction in expenditure, he allocated the benefit of the increased warmth in a range from 0 per cent to 100 per cent of this, according to whether people had previously been warm or not (1985, pp.58-9): either the whole of the £5 was saved or the whole of it was taken as additional warmth. This provides the maximum range within which the benefit occurred and, based on the research findings quoted above, will be artificially wide. He was not able to apportion numbers of households into the different categories.

The same project would obtain quite different results from these different methods, even excluding the fuel cost parameter. There has been no consistency of approach, partly because there is no consensus on method and objectives, and partly because of the variety of data collected. In none of the projects considered is there even a discussion of the limitations or advantages of the method of evaluation chosen. The mechanics of how to value warmth, and therefore which data to collect, are in need of clarification and definition. Only when this has been agreed on can data consistency be achieved and overall trends and relationships observed. Meanwhile, there is a risk that the assessment of the energy savings from demonstration schemes and the benefits of replication are being overvalued.

None of the scenarios cited exactly matches Barnett's model, partly because of the emphasis on energy units, though B vs. D is the nearest. If the temperature is kept constant, despite energy efficiency alterations, the consumer surplus is represented by the rectangular shaded area (Figure 9.2). Taking Plymouth as an example (Figure 9.5), if the household had been able to achieve the higher temperature of 17°C in the old house, the total benefit of the energy efficiency improvements would have been greater by the size of the triangle ONX — an additional £14, above the present total of £108. Because the Plymouth homes were not adequately warm, prior to the energy efficiency improvements, the total benefit is lower by 11 per cent, as measured by this method. The size of this triangle OXN is always relatively small. Thus:

— at any given level of energy efficiency, the greatest benefit (increase in consumer surplus) occurs in warm homes;
— improvements to a cold, inefficient home are often more cost effective than investment in a warm, energy efficient one;
— the colder the home originally, the more benefit is taken as additional warmth, rather than fuel savings.

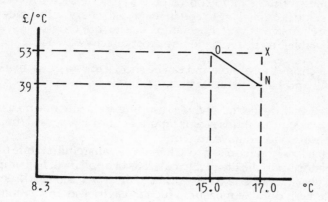

Figure 9.5 The effect of original temperature on the benefit derived from energy efficiency improvements at Plymouth, Better Insulated House (BIH) Programme

Note: The costs include delivered energy purchases and standing charges

Source: Based on Campbell 1985

The final problem is to indicate how this analysis can be made into a predictive tool for the purposes of assessing policies. Until new software can be written to model this analysis, it is proposed that a house is assumed to be adequately warm prior to any improvements, and the benefit assessed on this basis. It is up to the individual household to decide what proportion of this benefit they wish to take as additional warmth, and how much as reduced expenditure. This means that the cost benefit analysis is based on the B vs. D scenario and slightly overestimates the benefit obtained in previously cold homes. This comparison is a standard part of most computer energy audits, though it is normally used in assessing the cost effectiveness of design adjustments in proposed buildings. The importance of this economic analysis is that there is little loss in accuracy if the present temperature and expenditure are ignored. Therefore, through economic analysis, the way to value warmth and the necessary software application, has been indicated.

Cost-effective energy efficiency improvements

The cost effectiveness of a proposed measure depends on the capital costs involved, the price of the fuel originally being used and the rate of return required. Most of the concern here is with government criteria for cost effectiveness, because the poor do not have capital themselves. The government requires public sector investments now to achieve a minimum 8 per cent real rate of return; previously it was 5 per cent. Thus, officially, the government are expecting public sector investments to achieve a return that is more comparable to private sector investments. However, some fudging is going on in relation to energy efficiency policy. In a comprehensive report on energy use in buildings, the Building Research Establishment used the term cost effective to mean:

> what is technically possible and can be shown to be a good investment. (Shorrock and Henderson 1990, p.11)

It is accepted that the actual savings resulting from any one measure can be building specific, but this seems an unnecessarily vague definition on which to base advice to government.

The payback period required to achieve a certain required rate of return (RRR) depends upon the life of the materials or appliance. For simplicity a 10-year payback can be taken to provide an 8 per cent RRR; longer is permissible with durable measures such as cavity wall insulation. Using figures provided by the EEO (1986) at November 1985 prices, for 13 house types, different fuels and a variety of insulation measures, the payback on various options has been calculated (Table 9.5). The effect of the fuel is clear, with the columns arranged in cost order, and measures in sequence of cost effectiveness.

Although these figures are based on whole house heating, the EEO

Table 9.5 Payback periods for energy efficiency measures in relation to the fuel used for heating (years)

	On-peak elec.	Solid fuel	Off-peak elec.	Gas
Cylinder cover	<1	<1	<1	<1
Loft insulation	<1	1.5-2.1	1.7-2.2	2.3-3.2
Draughtproofing (1)	<1-1.7	2.0-4.0	2.0-4.0	3.0-4.0†
Cavity wall fill (2)	1.1-2.4	2.3-5.5	2.5-5.5	3.2-7.3
Solid wall — internal	3.4-7.7	7.0-11.0†	7.9-12.2†	9.7-10.8*
Solid wall — external	7.7-11.5	12.2*	na	na
Double glazing (3)	na	na	na	na

Notes: The cost of the measure and resultant savings vary with the shape of the external envelope, as well as the cost of the fuel used; average costs are given in Table 10.3; the range of figures is for different house types, e.g. flat or detached house; all day (16 hour) heating assumed, to meet the needs of low-income families; work to be done by a builder, not DIY.
(1) with a life of at least 6 years;
(2) based on most expensive form of insulation — mineral fibre;
(3) double or secondary glazing is only cost effective, if expenditure is below £700 and the product lasts 20 years;
† achieves 5% real rate of return in most, but not all, property types;
* only cost effective in older, terraced houses;
na not applicable, as the payback period is longer than 12.5 years

Source: Based on EEO 1986

assumes that the living room is heated to an average of 21°C for 16 hours a day, but does not stipulate what temperature is wanted, or achieved, elsewhere. Only double glazing is uniformly not cost effective, and all the other measures represent extremely good investments in many homes. The government's emphasis has rightly been on loft insulation and draughtproofing, though the omission of any policies to encourage the use of cavity wall insulation is not based on the potential benefits.

Similar calculations have been carried out for energy efficiency improvements through the replacement of new heating systems for old ones, including, if appropriate, fuel substitution. If the life of the new system is assumed to be 20 years and the present system is one of the least efficient, the amounts of capital expenditure that can be justified are usually extremely generous, particularly if some of the old system is retained (Table 9.6). Replacing the systems in the 1.8m low-income homes dependent upon on-peak electricity for heating or expensive CH would be extremely cost effective. Thus, in the most energy inefficient homes, capital investment in many measures to improve the energy efficiency is cost effective on the basis of accepted public sector investment criteria. Appropriate fuel substitution is the most effective energy efficiency measure for many low-income families, because it influences all heating costs (not just a proportion of the heat loss as with insulation measures).

Table 9.6 Capital expenditure limits when changing fuel systems (£ per house)

	Old system			
	On-peak elec	Solid fuel	Off-peak elec	Gas
New system:				
Gas CH	<14,500	600-1,900	1,300-4,000	500-1,400
Solid fuel CH	<15,000	yes *	1,500-4,300	600-1,850
Night storage heaters	<13,000	yes *	700-2,500	na

Notes: The adequacy of these sums of money to purchase a new CH system varies with the size of the dwelling, the amount of the previous system that can be utilized and the rate of heat loss. The range of figures represents the capital available in different house types, e.g. a flat or detached house. The smallest sums are for flats, where it is estimated that the money is sufficient to buy at least three outlets which may constitute partial or full CH. In most cases, it is anticipated that the money available is adequate to provide at least partial CH. New systems are assumed to have a life of 20 years. The estimates are based on an all day (16 hours) heating regime, to meet the needs of low-income families; work to be done by a contractor, not DIY.
* very cost effective if replacing open coal fires. Less likely to be cost effective otherwise, particularly in smoke control zones;
na not applicable — either no or negligible savings in running costs.

Source: Based on EEO 1986

In the past government policies have focused on the addition of insulation to the building fabric and little attention has been paid to the energy efficiency of the heating system and the benefits of grant-aiding fuel substitution. Government advisers continue to be reticent about the benefits of fuel substitution and a recent publication still does not list fuel substitution as one of the twelve most cost-effective measures, though it does include 'full double glazing in all homes' (Henderson and Shorrock 1990, p.3). This is using the definition of cost effective cited above, from the parent publication (Shorrock and Henderson 1990). No solid wall insulation is listed as cost effective. The logic behind such a ranking is impossible to understand, particularly as it is linked to energy saving. Measures such as double glazing become cost effective if they are undertaken when more general repairs are needed and the window needed replacing. The extra cost of putting in double rather than single glazing can be negligible. In other cases, a measure will be seen as having other benefits, such as reducing maintenance or noise pollution, and the energy saving becomes only one of the criteria.

The control of greenhouse gases is becoming another reason for reducing energy consumption and is adding important new investment criteria. The greatest attention is being paid to the reduction in carbon dioxide (CO_2) emissions. The different domestic fuels vary by a factor of four both in their cost (Table 5.1) and in the amounts of CO_2 (Table 5.3) that they produce. Policies that reduce both the cost of energy services and carbon dioxide emissions are expected to come to the fore, so that there should be a greater emphasis on fuel substitution approaches in future. Jackson and Roberts have shown that fuel switching (in domestic and other sectors) is the single most cost-effective way to reduce carbon dioxide emissions and results in greater benefit than improved appliance efficiency (1989, p.11). By the summer of 1990, no policy proposals had been made by government.

Investment criteria

Government

Before the privatization of gas and electricity it was possible to make comparisons between the investment by government in the more efficient use of energy and the provision of additional supply. The Secretary of State for Energy had stated that they 'ought both to be determined according to similar criteria' (DEn 1983a, p.2), though this did not happen in practice. Many cost-effective investments in low-income homes were not undertaken. One reason was muddled thinking at the DEn.

In a DEn document (1983b) on investment in energy use and energy demand there was a failure to differentiate between the demand for energy and the demand for warmth. With reference to the substitution effect of price elasticities:

> As the cost of keeping his house at any temperature falls, the consumer tends to consume more energy than before.(ibid, p.21)

The discussion earlier in this chapter and Figure 9.1 have shown that as the cost of warmth falls, the demand for warmth increases, but average consumption, even in fairly low-income homes, falls. Perhaps because of the inadequacies of the DEn analysis (1983b), the Select Committee on Energy continued to request a thorough comparison of the benefits of investment in energy supply and energy use (HC 262 1986, para 8). With the privatization of the electricity and gas industries, the comparison between investment in energy supply and demand is no longer an appropriate requirement of government.

The quote from DEN 3 above (p.186-7) is unequivocal about investment in public sector housing: if the net benefits are large enough, it does not matter who actually receives them. Therefore, the same investment should take place whether the property is in the public sector, owner occupied or in the private rented sector. As householders value additional warmth, so should the government, even where the beneficiary has obtained the improvement for free, as a result of government grants. This is the approach taken by the Property Services Agency (PSA), on behalf of the Ministry of Defence: energy efficiency improvements to the homes of married personnel are paid for by the PSA, provided they 'demonstrate a pay back period of 10 years or less to satisfy Treasury investment criteria', even though the benefit of lower fuel bills and extra warmth accrues to the occupants (Hammond 7 Mar. 1986, pers comm).

The policy of the EEO, as stated to the Select Committee on Energy, appears to be less positive. The concern about investing in the homes of low-income groups is:

> how much of the improvement in energy efficiency which might result … will be taken as increased comfort standards and how much will result in lower energy bills. That is a real problem because it makes it difficult to argue it from a public finance point of view. (Macintyre 1985, p.28)

This explanation implied that public expenditure can only be justified when the return is achieved from energy savings alone and that additional warmth is not included in the value of benefits. Even more stringently, an investment is only worthwhile if there are resultant savings to the Exchequer. This is a different perspective to that of the DOE and the PSA, both of whom recognize that all benefits are of importance, no matter who they accrue to, and are not concerned solely with government savings. These policy differences are disturbing, as it is obviously important that all departments operate with the same investment criteria, particularly in the same area.

More recently, the need to assist low-income households has been recognized by the EEO:

> many of the households most in need of improved energy efficiency

are unable to afford the measures that would enable them to obtain better value from their energy expenditures. (HC 262 1986, p.xiii)

Therefore, the EEO aim:

to accelerate the installation of energy efficiency measures in low-income households. (HC 262 1986, p.xiii)

As stated in Chapter 4, no new policies were developed by the EEO, beyond support, with other departments, for community insulation groups. The targeting of loft insulation grants on the poor was a DOE initiative. The proposed HEES is due to take effect from the beginning of 1991, but could in practice be the same policies wrapped in new names. For the 'acceleration' of installation there has to be a level of investment two or three times greater than that being suggested. Even the proposed level of involvement is inconsistent with the belief of the then Secretary of State for Energy, Peter Walker, that fuel poverty is no different from poverty and can be reduced through income support. Therefore, it can be to the benefit of the poor that the government is inconsistent. The mismatch between statements and actions confirms the view that the real causes of fuel poverty are poorly understood. The Select Committee on Energy has criticized the government for its lack of policies on low-income households consistently since 1982 (most recently in HC 87 1985, p.vi) and similar comments were made in the Rayner Review (Finer 1982).

Part of the confusion may stem from what is defined as the 'public sector' and therefore what is included in the domain of public sector investments. There are three different situations:

— the state owns the building and pays the fuel bills, e.g. prisons;
— the state owns the building, but does not pay the bills, e.g. forces' housing;
— the state does not own the building, but pays the fuel bills because the occupant is dependent upon benefits.

Even if it is clear what is meant by the public sector, there remains the issue of departmental responsibility. With the example of the Ministry of Defence, it is the landlord department that is investing in energy efficiency. If the same principle is applied to local authority housing, the DOE would be responsible. Investment in the homes of low-income families dependent upon benefits but living in other tenures would come from the EEO or the DOE. The lack of clear departmental responsibilities, highlighted in Chapter 2, is one of the reasons for uncertainty about the proper policy locus. If one government department (the DSS) is ceasing to give money towards heating costs, the logical inference would be that there is a greater need for capital investment by the EEO or DOE, not that there is less justification for this expenditure.

This morass of confused and conflicting statements and criteria, particularly from the DEn, demonstrates the need for a consistent basis for assessing public investment programmes in energy efficiency improvements to low-income households and the following are suggested:

— the value of additional warmth is included, as well as the present energy savings. For simplicity, the house is assumed to have been adequately warm beforehand and the benefits assessed on this basis;

— all benefits are included in the economic analysis, regardless of whether they accrue to the occupant, government or some other person;

— the public sector includes buildings owned by the state and the homes of those households dependent upon benefits for at least 75 per cent of their income;

— the desirability of the improvements is not affected by whether or not there are any savings in other government programmes.

At the moment, the DEn's policy on energy demand reductions, particularly in low-income households, is the main aberration. In other decisions on energy efficiency measures outside of the DEn, the procedure for a proper economic analysis is followed. Therefore, investment in energy efficiency improvements to the homes of low-income households would pass the Treasury's rate of return requirements if assessed by the DOE, but might not even be considered by the DEn. The strong influence of other policy considerations demonstrates that investment criteria are not, and cannot be, applied in a consistent manner either within or across departments.

Despite considerable discussion, no target for carbon dioxide emissions has been agreed to or set by the British Government by summer 1990. There are tentative suggestions that despite some expected rate of growth, we might hold our pollution level at the 1985 level. As this section has shown, there are numerous cost-effective energy efficiency measures available that would meet all government criteria. A reduction of 25 per cent of emissions from the domestic sector is certainly possible and up to 35 per cent is technically possible (Henderson and Shorrock 1990, p.3).

Privatized utilities

The effect of privatization is that whilst supply investments are the responsibility of the supply companies, they have not a comparable responsibility for investment in the efficient use of energy. There is, therefore, no forum in which a direct comparison between the two types of investment can take place. This inevitably means that cost-effective investment in energy use will not take place, whereas additional new supply will be funded.

The comparison is needed, because energy supply and energy demand are competing for that scarce resource, capital. In both cases, large amounts of money are involved, certainly £billions, and the nation will benefit most from the investment that gives the greatest return. There is no intrinsic difference between the cost benefit analyses required by energy supply and energy demand and in both cases the main beneficiary is the consumer through increased energy efficiency: in one case the investment results in a lower cost of delivered energy (for all consumers), in the other the effect is to lower the cost of warmth (for domestic consumers).

Unfortunately, despite the best efforts of the House of Lords, both electricity and gas privatization Acts limit the responsibilities of the industries towards the efficient use of energy. In the Gas Act, section 4 requires the Secretary of State for Energy and the Director of Ofgas (the gas Regulator) in the exercise of their functions:

> to promote efficiency and economy on the part of persons authorised … to supply gas through pipes and the efficient use of gas supplied through pipes.

There are several problems with this. First, there is no comparable requirement on the industry itself, only a condition in the Licence to Supply, which requires the preparation of a 'statement' giving 'general information for the guidance of tariff customers in the efficient use of gas'. This statement was prepared in November 1986, has not been widely circulated and is now out of date, for instance because of references to Supplementary Benefit. Secondly, the Director General of Ofgas has relatively few 'functions', so that his opportunities for intervention are extremely limited and occur mainly when setting the pricing formulae. Thirdly, even when he can intervene, he can only require the suppliers to 'promote' the efficient use of gas; he cannot, for instance, ensure that they invest directly in low-income homes.

The legislative framework for the efficient use of electricity is exactly the same as for gas, with one important addition. A new clause was added to the Act at final reading which permits (not requires) the Director to:

> determine such standards of performance in connection with the promotion of the efficient use of electricity by consumers as, in his opinion, ought to be achieved by such suppliers, (section 41)

and arrange for the publication of these standards. Different standards can apply to different public electricity suppliers (PES — the old area boards). This new clause still refers to 'promotion', so that the net effect may be nil for low income consumers in need of investment. All of the legislative problems outlined for gas, therefore appear to apply to electricity.

There is a further constraint on the supply industries. The costs of both distribution and supply that can be passed on to consumers are tightly defined by formulae. The cost of any investment in the more efficient use of energy is not allowed to be passed on to customers through these formulae. Therefore, if a supplier made such an investment, it would have to absorb the costs internally whilst at the same time suffering the cost penalties of reduced demand. Both gas and electricity industries have been privatized to function solely as suppliers of energy, just when society needs investment in supply and demand to be more closely integrated.

Integrated energy planning occurs in other countries, particularly the

USA, and is usually known as least-cost planning. This focuses on whether it is most cost effective to provide energy services through additional supply or improved energy efficiency (demand reduction). The issues involved, with special reference to British Gas and the first tariff review, have been admirably covered by Brown (1990). As he states, consumers are concerned about the size of their bills and the amount of energy services they have received for that expenditure. They are not so concerned with the tariff of the fuel purchased. In addition, the USA is working towards energy least-cost planning, that is comparisons between both supply and demand management across a range of fuels, not just for a specific fuel. The UK has moved in the wrong direction at the wrong time.

Conclusions

The cost of combating fuel poverty through increased income support would be about £15bn in 1988 in revenue expenditure alone, rising in line with inflation thereafter and any growth in the numbers of eligible recipients. Undoubtedly, this level of support would be politically unacceptable to all administrations, as it is equivalent to an additional 9 per cent of total public expenditure.

The development of a policy to achieve cost-effective capital investment in the energy efficiency of low-income households is hampered by the sublimation of consistent investment criteria to other objectives, particularly restraint on local authority expenditure. In the past, one obstacle has been the lack of a clear procedure for valuing warmth. This is no longer a problem because Barnett's economic model has been developed, through the use of research data, into a method that can be used consistently and with ease with energy auditing computer programs. At a macroeconomic level, the debate about the cost-effective use of resources in energy efficiency measures can easily get excluded by the important, parallel assessment of investment in adequate supply. The value of warmth is an important concept at both household and national level.

The use of economic models and analysis have provided the rationale for observed behaviour as well as a framework for evaluating policy options. Low-income households demonstrate by their high income elasticities that they do suffer from fuel poverty and will use additional income to purchase extra energy and warmth. This is especially true of 81 per cent of single pensioners, 51 per cent of single-parent families and the 18 per cent of households known to be spending a disproportionate amount on fuel already. It has not been possible to determine the price elasticity of demand for warmth in low-income households, but in the evidence examined, higher temperatures are closely linked to greater energy efficiency. This confirms the finding in Chapter 7 where temperatures are higher in younger dwellings. Thus, cold families will have warmer homes if either their houses are made more energy efficient or they receive additional income: in the former case there are energy savings, whereas energy demand increases in the latter situation. Thus, for the poor to be warmer involves capital expenditure either on their homes or on the provision of additional supply.

In response to a fuel price increase, a significant number of low-income families would decrease consumption apparently because of the low marginal utility they obtain from additional purchases of energy. This is not because the people in these families prefer to live in colder homes: as has been shown in Chapter 6, similar temperatures provide comfort conditions for nearly everyone in the population. It is the cost of warmth that needs to be lowered for these families. There is, therefore, a need to distinguish carefully between the demand for energy and the demand for warmth.

10 A programme for affordable warmth

To resumé the findings so far: there are substantial variations in the energy efficiency of houses, to the detriment of the poor, so that they are unable to obtain enough warmth for either health or comfort. To enable the poor to purchase as much fuel as other households, through income support, would cost at least an extra £15bn annually, indefinitely. This hugely expensive policy would be subsidizing energy inefficiency and would still not guarantee adequate warmth. The additional energy demand generated through this extra income support could only be met through substantial investment in increased energy supply and large increases in carbon dioxide emissions. Thus, a government policy to overcome fuel poverty through revenue expenditure would result in capital expenditure by the supply industries, an increase in fuel prices and a subsequent worsening of fuel poverty again: an expensive, polluting and spiralling approach.

A programme for affordable warmth is proposed, together with some of the detailed policies that would have to accompany it. The emphasis however is on the broad objectives of any scheme designed to reduce hardship for the fuel poor. Underlying this scenario is the assumption that the cost of keeping warm in a house can be defined, together with the amount of money available for heating within the family budget. Both of these elements need to be clarified and understood if 'affordable warmth' is to be made more widely accessible.

Defining fuel poverty

Energy labels

Recognition of the wide variations in the energy efficiency of dwellings underlies all the proposals in this book. To make most impact on fuel poverty these variations in energy efficiency need to be at least identified and probably quantified. The first stage is to clarify the energy standard required — extending the recommended heating standard — to ensure that adequate warmth is accompanied by sufficient other energy services, such as hot water and electricity for appliances. Then the energy standard can be converted into an energy label (or energy code) for each house through an energy audit. The final label needs to reflect the costs involved, because of the wide variation in fuel prices, and include fixed costs such as standing charges and maintenance of the household's heating equipment. If energy labels are to be used in rent assessment and as the basis for additional income, as proposed here, it would be easier if there is a close relationship between the energy code and household costs.

An energy audit of a building can be undertaken in a variety of ways with varying levels of skill, time and cost. Most reflect expected expenditure by the household, rather than actual fuel bills, which is appropriate. Some of the old HA were based on energy efficiency criteria, for instance the cost of the fuel (estate rate HA). These applied one or two values to a single parameter: the simplest and crudest form of energy efficiency rating, as it is basically a 'yes' or 'no' coding. At the other extreme is a computer-based analysis, generally costing in the range of £50-£200 per house. Most, if not all, variations on this approach are based on BREDEM (see Glossary) and measure all energy use, usually converting annual energy expenditure into an index (MKECI), a rating (the embryonic National Home Energy Rating scheme — NHER), or points (Starpoint). The holistic approach to domestic energy use, based on BREDEM, is an increasingly important part of the domestic housing scene in Britain and, once an appropriate standard of energy service has been defined, could be of considerable benefit to the poor. The use of energy labels for buildings has been advocated by advisers to the Secretary of State for Energy (HC 87 1985), but the government has provided only luke-warm support (HC 405-i, 1990, p.6).

What is affordable?

The negative definition of fuel poverty — the inability to afford adequate warmth — has been accepted for some time, but there have been no attempts at clarifying what is 'affordable' warmth. Whilst energy audits provide a method of measuring the cost of warmth, or all energy services, there is no basis for deciding whether this represents an affordable amount of money for the household. There are three ways in which household expenditure on all energy (including standing charges and maintenance of heating equipment) could be identified:

— existing expenditure: this is time consuming because it requires a detailed estimation of each household's fuel bills;
— each household is assumed to have the same sum of money for energy regardless of details of the accommodation and number of people occupying it. This is the NFE (see Glossary) approach used originally with SB claimants when the rent includes heating costs and which is being phased out with the change to IS. The arbitrariness of applying the same fixed sum of money to all households is not satisfactory, as small households have to pay a large proportion of their income before receiving additional help;
— the defined level of service is affordable, providing that it does not exceed a certain percentage of the family's income. The rationale here is the same as that behind HB: there is a limit to the amount of

money a low-income family should be expected to pay for adequate shelter, whether through rent or fuel costs.

In all three situations above, if the energy label demonstrates that the cost is greater than the money the household has available, additional assistance will need to be given.

The most equitable way of assessing a reasonable amount of expenditure for fuel is judged to be as a percentage of income. Total income is already assessed for all IS and HB claimants, so that no additional means-testing should be required, at least for the majority of low-income households. Deciding what proportion represents a 'reasonable' or 'normal' level of expenditure is problematic.

As shown in Table 3.4, the poor spend twice as much, as a proportion of income, as the rest of the population, though this is less in absolute terms. The difficult question is whether low-income households should be able to purchase adequate energy for a smaller proportion of income than they at present spend: that is for less than 10 per cent of total expenditure. As low-income families are likely to be at home most of the day, they could be expected to spend more on heating than those households where most people are at work, or out. However, as Table 6.4 showed, the increase in the numbers of room-hours of heating required because of greater hours of occupancy is virtually offset by the larger number of people in better-off families. To propose that energy expenditure should be the same proportion in all families is, effectively, to say that there should be parity of income levels. Expenditure on a basic necessity, such as fuel, could only be reduced to 5 per cent of a poor family's income if their total income was at least doubled. This is why the additional income support route is so expensive to implement. Therefore, 10 per cent of income is taken as the amount that is 'affordable', though the actual figure does not effect the description of the scheme, only its detailed administration and costs. (On a pro rata basis, adequate heating is affordable if it can be obtained for 6 per cent of income, because 60 per cent of the average household's energy consumption is for heating.) Therefore, it is proposed that if a household's energy costs are above 10 per cent of their income, they should be entitled to receive additional assistance. This will mean that, for the same expenditure as at present, they will have an adequate level of energy services, including warmth. To this extent, they will be better off.

One of the benefits of defining reasonable energy expenditure in relation to total income is that the amount of money available generally increases with the size of the family. As there is, generally, an equivalent association between the number of occupants and the amount of space in the dwelling (Table 6.6), these two relationships should combine neatly. Affordable warmth (and other energy services) can be expressed as an amount of money per unit of floor area, as in the MKECI: large families have both more money for heating and additional space to keep warm.

For policy purposes, a simple method is needed to relate household income, affordable warmth and dwelling size. Two possible measures have been plotted (Figure 10.1): a MKECI of 120 and a standard proposed

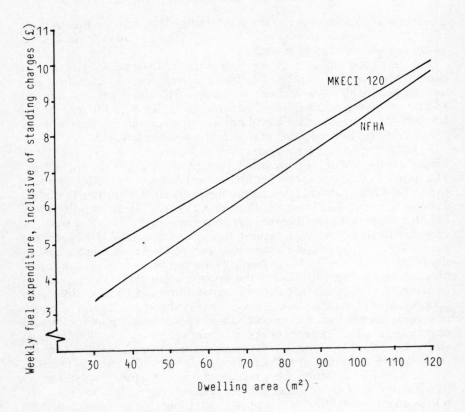

Figure 10.1 Relationship between fuel expenditure and dwelling area using two standards for affordable warmth

Notes: Weekly fuel expenditure is the total the household have available, based on 10 per cent of their income. As the price of fuel varies, the mix of fuels available to a particular household determines the quantity of energy they can afford to purchase: households using expensive fuels need higher insulation levels to compensate.
MKECI 120 includes standing charges and maintenance, but not depreciation. The formula includes a variable to reflect changing fuel costs; 39 was used in these calculations
NFHA interim suggestion of 7p/m^2 week excludes standing charges, maintenance costs and depreciation, though standing charges have been added back in to obtain weekly expenditure; 1987 prices.

Source: Author's own calculations

by the NFHA's Working Party on Energy Conservation and Fuel Poverty:

— the MKECI is designed so that houses with the same level of energy efficiency, per unit of floor space, have the same index. A minimum of 120 was originally set for the Milton Keynes Energy Park, still considerably better than the 150-160 standard of the 1990 Building Regulations. The MKECI includes standing charges and heating equipment maintenance;

— the NFHA's interim suggestion for affordable warmth is a direct measure of all energy services, not an index. It is therefore unique and a particularly useful approach. The total cost of providing the required standard, exclusive of standing charges, should be no more than 7p (1987 prices) per square metre of floor area, per week. When this is calculated as total fuel expenditure per week (and the standing charge is added back in, though there is no allowance for maintenance costs) the result is a line with a steeper gradient than the MKECI 120. This is better for occupants, as running costs are lower, but more difficult for building professionals to achieve.

From Figure 10.1, a household with a total weekly income of £70 and with £7 (10 per cent) for fuel would have affordable warmth (and all other household energy services) if they lived in a dwelling either smaller than

Table 10.1 Relationship between income and dwelling size using two standards for affordable warmth for specific groups, UK 1985

Household type	10% of income for fuel (£)	Maximum space on present income	
		MKECI 120 (m²)	NFHA (m²)
Low income* pensioners, 1 or 2 people	4.65	30	48
Single pensioner in HA** 1 bedroom unit	4.78	32	50
Single parent family in HA** 2 bedroom unit	5.68	47	62
Low income* single parent family	6.68	64	77
All households in HA** 2 bedroom units	7.40	76	87
All households in HA** 3 bedroom units	9.67	112	118

Notes: Data for the last two columns are from Figure 1, ie a MKECI of 120 and the NFHA standard of 7p/m²/week
* 20% of households with lowest incomes
** housing association or housing co-operative

Sources: Family Expenditure Survey 1985; Figure 10.1

$70m^2$ and a MKECI of 120, or than $81m^2$ using the NFHA measure. Data from the Family Expenditure Survey 1985 and the NFHA New Lettings Survey (1985) provide some examples of specific family types and the maximum space they could occupy without further financial assistance (Table 10.1). If additional money is not available, the minimum standard of heating will not be achievable for 10 per cent of income, and hardship occurs.

Until further data are obtained, it is not possible to decide whether either of these standards is an appropriate definition of affordable warmth for low-income households. The 120 MKECI, as at present formulated, does not appear to be rigorous enough as it results in too little space for small families. For instance, the minimum standard for a self-contained unit for two people is probably about $45m^2$ (based on the recommendations in the Parker Morris Report — MOHLG 1961). The steeper gradient of the NFHA measure provides a more generous space allocation for households on the very lowest incomes. However, the MKECI is used in the remaining discussion, because it is more widely recognized. A 120 MKECI is equivalent to a rating of 8 on the NHER scheme, whereas the 1990 Building Regulation standard is about 7, out of a maximum of 10 for extremely energy efficient houses.

In order to decide on a minimum cost index, reference is made to the work of the economists Isherwood and Hancock (1979) cited in Chapter 3. They used multiples of median expenditure on fuel to define disproportionate expenditure, and by implication hardship. If the same principle is applied, the following MKECI are achieved, based on median fuel expenditure in the whole population. This varies slightly from year to year, but 6 per cent of income is taken as normal average. These guidelines are only applicable for low-income groups, because the same proportion of income results in greater absolute expenditure in better-off families, and thus different MKECI levels are affordable.

120 — below this level fuel poverty probably does not exist, as adequate energy can be purchased for about 10 per cent of income. It is only low-income families in homes indexed below this level — the target for all low-income homes eventually — that do not need additional income support;

160 — a low-income family in this grade house is spending about twice the median of the whole population: 12 per cent of income on fuel;

275 — three times the median: 18 per cent of income on fuel;

400 — four times the median: 24 per cent of income on fuel. Above this level, it is suggested that the house is unfit for human habitation: it is too expensive to keep adequately warm and thus does not provide sufficient shelter. The occupants of a dwelling with an index above this level are at considerable risk of cold-related ill health.

It is worth reiterating the comments quoted earlier about the rationale behind HB: according to the Treasury, housing costs have to be treated separately because the cost of similar accommodation could vary by a factor of two across the country. The size of the variation, in practice, between energy costs in similar dwellings cannot be quoted here because of the lack of data (though some evidence is given below). A factor of five was established for Birmingham's local authority housing by Byrd — though this range could not be linked to dwelling or family type — and a factor of three with the Better Insulated Houses (Figure 7.3). Homes that are less energy efficient than the 120 MKECI, and have higher codes, cannot provide an adequate standard of energy services for a low-income family for 10 per cent of their income. This is the first quantitative definition of fuel poverty.

It is now possible to redefine fuel poverty: it occurs when a family are unable to afford adequate warmth because they live in an energy inefficient home.

Extremes of fuel poverty

Fuel poverty can now be quantified, but existing data are not detailed enough to estimate the numbers of households affected. It is only possible to give examples of families at the two extremes: suffering great fuel poverty and least likely to have difficulty obtaining affordable warmth.

Consumption costs have been obtained from a sample of audits of local authority flats both as they are now and as they would be after two different levels of capital expenditure (Table 10.2) to demonstrate how great fuel poverty can be. It is not suggested that these are the most

Table 10.2 Total energy costs in existing local authority dwellings

Estate (Floor area)	BEFORE Running costs* (£/w)	AFTER Capital expenditure (£ per unit)	Running costs* (£/w)
Debdale House (41m²)	14.70	1700-2400	4.30-3.50
Wenlake House (53m²)	18.30	2000-4500	5.50-4.00
Bracklyn Court (58m²)	19.50	2050-4900	5.60-3.20
Stanway Court (60m²)	16.20	1800-2100	4.00-3.40
Elthorne Estate (74m²)	8.70**	2000-2800	6.90-6.00

Notes: These blocks have been chosen at random from audits undertaken by ECSC. The costs are an average of those to be found in each block and are based on a temperature of 21°C in living rooms and 18°C in other rooms for 16 hours per day, 7 days per week. 1987 prices.
* includes 2.44p/w/m² for electricity for lights and appliances based on the standards in Energy Designer software and Manual;
** district heating costs subsidised by the local authority at present.

Source: Energy Conservation and Solar Centre 25 June, 1987, pers comm

extreme examples possible, merely that they demonstrate the known problems from a relatively small sample of properly audited dwellings. There is no information about the income levels of the households that occupy these properties, nor, therefore, what represents affordable energy expenditure. The range of capital expenditure shown is all cost effective and gives payback periods of 2-8 years, except for Elthorne, which cannot be calculated accurately because of the local authority subsidy. As a result of the improvements, running costs are reduced to a third or a quarter for the same standard of energy services. The cost of these, per unit of floor area, is highest in the smallest flats.

As two further examples of existing standards, properties in one 1930s block of flats had MKECI ratings of 648-760 before improvement and contained serious condensation problems as a result of being hard to heat (Burton and Hughes 1985, p17). An expenditure of £4,600 per flat is needed to provide affordable warmth and lower carbon dioxide emissions in small, 4-person, Glasgow flats (Boardman 1990, p9). To achieve an adequate standard of energy services would have cost £22 per week beforehand, but £8.50 after the capital improvements. It is not known what proportion of dwellings occupied by low-income households are this energy inefficient. However, because it is better than current building regulations, the affordable warmth standard (120 MKECI or similar) represents an extremely large improvement in the energy efficiency of most British homes, particularly those occupied by low-income groups.

At the other end of the spectrum there are those low-income households who are not suffering from fuel poverty, or only slightly. Only a small number of homes in Britain are likely to be able to obtain a 120 MKECI or less — perhaps less than a thousand, for all income groups, and these are mainly on the Milton Keynes Energy Park. Some poor households come close to the standard, for instance the 2 per cent of DE households known to have cavity wall insulation, double glazing and loft insulation (Table 4.4), particularly if they also have gas CH. Therefore, fuel poverty and poverty apply to virtually identical groups. To be conservative, of the 6.7m households defined as in poverty in 1990, all but a handful (perhaps 100,000) suffer from fuel poverty. Because the overlap between poverty and fuel poverty is now shown to be almost complete, it is possible to draw and label the Venn diagram first introduced in Chapter 3 (Figure 10.2).

There is still some imprecision, because, for instance, of uncertainty over the number of families that are dependent upon the state for at least 75 per cent of their income and occupy their own home and thus the extent to which they are the same population as people in receipt of HB. Another unknown is the effect of non-claimants — as stated, about 2m households are thought to be eligible for HB and not claiming. This diagram indicates the likely overlap between the groups, therefore, and is not a precise representation. However, at least 30 per cent of all families in the UK suffer from fuel poverty. In the remaining calculations, 6.6m

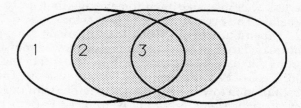

Figure 10.2 Venn diagram to illustrate the relationship between fuel poverty and two definitions of poverty

Notes:
1 — HB recipients 5.5m+ (see Chapter 3)
2 — Fuel poverty suffers 6.6m
3 — Household receiving at least 75% of its income from the state 6.7m

households affected by fuel poverty is the number used for policy targeting. All of these households are in receipt of a means-tested benefit — they have been defined as poor by the government's own criteria — and are unable to obtain an adequate level of energy services, particularly warmth, for 10 per cent of their income.

The proposed programme

There are 6.6m poor families living in homes that are too energy inefficient for them to heat adequately. All of these homes need a substantial package of insulation measures and improvements to the heating system if they are to provide affordable warmth. These measures could be provided separately, as with loft insulation and draughtproofing grants, or they could be included in a comprehensive Programme for Affordable Warmth (PAW).

The standard to be achieved is the 120 MKECI, or equivalent, and this is approximately the thermal insulation levels required by the 1990 Building Regulations (Table 4.1) together with double glazing and gas CH. Where gas is not available or would be unsafe then a higher level of insulation would be needed to compensate, for instance more than 150mm of loft insulation or triple glazing.

Despite uncertainty about the dimensions of the average low-income house, the amount of existing insulation and the opportunities for bulk contracts, an average cost per dwelling has been taken to be £2,500. This is a combination of £1,400 to bring the average house up to the 1990 Building Regulation standards for thermal insulation (based on Bowdidge, p.13), with a reduction for the cost of cavity wall insulation. Bowdidge quotes a figure of £7/m², whereas the South London Consortium can obtain £2/m² through bulk contracts, and this lower figure has been used. An allowance of £1,100 has been included to treat the windows, heating system and controls. This may be an underestimate, because low-income homes are less well insulated than the average, but, in compensation, they are smaller. The average figure of £2,500 for all the 6.6m dwellings gives a total expenditure of £16,500m. In comparison, the proposed rate of expenditure on HEES of £12m (including agency administration costs) for six months will barely keep pace with the growth of the problem.

The major focus of PAW is to improve heating systems. The slower rate of fuel substitution in low-income homes together with the expense and inadequacy of the systems presently being used is probably the main cause of fuel poverty. Some households could not be connected because they are too far from a gas main (15 per cent) or because they live in a block of flats where conversion would be unsafe (probably a relatively small percentage). However, 12 per cent of all households are within 25m of a gas main but not connected and 750,000 low-income households have gas on the premises and do not use it for heating (based on Bradshaw and Hardman 1988, pp.9,13). Changing these families to a gas-fired system would be extremely cost effective, particularly if they are currently using on-peak electricity. The need for all rooms to be provided with a heating outlet, as recommended in the heating standard, might mean that an appropriate policy would be to upgrade heating systems to gas CH wherever practicable.

The reduced use of electricity for heating is vital both to reduce fuel poverty and to curtail carbon dioxide emissions. Policies and regulations should be developed for new and existing housing to ensure that substitutes are used for electricity where there is a less polluting alternative that is safe and acceptable to the user. This would reverse the recent growth in electric CH installations in new local authority dwellings (Figure 2.1) which can now be seen as an unfortunate development. In future, off-peak electricity should only be used to replace general tariff electricity, or where there is no alternative. This is the one area where traditional assistance for the fuel poor could conflict with environmental benefits. In the past, the wider use of off-peak electricity could be supported, because of its low price. Now, the use of gas has to be advocated wherever possible and Economy 7/White Meter systems are much further down the list, below efficient coal-burning appliances.

Global warming policies will have to focus attention on reducing the amount of electricity used in the home for all purposes, in addition to heating. Where alternative fuels can be used — with cooking and water heating — policies are also needed to prevent installation in new housing and enable fuel switching to occur in existing premises, wherever possible. The focus on electricity must extend to electricity-specific uses as well, such as lights and appliances. The emphasis should be on direct capital investment to obtain emission reductions in low-income homes, so that new forms of grant aid will be needed, as allowed under Section 15 of the Social Security Act 1990. For instance, American electricity utilities give low-energy light bulbs to low-income consumers as a cost-effective method of reducing demand.

The poor suffer from condensation and ill health as a result of living in a cold home. Whilst present concern is focused on the problems of excess carbon dioxide emissions and how to return the atmosphere to good health, the problems faced within the home by many people at the moment must not be forgotten. The standard being proposed would reduce fuel poverty and cut CO_2 emissions from the same households by at least 6 per cent to 6.0 tonnes p.a. per household (Table 8.8). This results from thermal insulation improvements and heating system substitution. Therefore, even the proposed PAW, which is limited to heating, produces benefits to both the poor and to the wider community through lower emissions and a healthier atmosphere as well as a cleaner, less polluted indoor environment.

To bring all low-income homes up to the standard of providing cleaner, affordable warmth will take a long time. The rate of new construction, particularly in the public sector, has declined, so most new housing is in the private sector and not accessible to the poor. As unwanted houses are being demolished at a rate of 12,000 p.a., it will take 2,000 years to replace the present housing stock. The majority of families will, therefore, only benefit from the renovation and upgrading of existing properties. The proposed standard is rarely achieved in improvements for low-income families. This is why it is possible to state that the vast majority of the poor are in fuel poverty — the lack of energy efficiency in British homes ensures that.

The year 2005 has been set as the target date for the completion of the programme. This date was chosen because it is the one mentioned in the Toronto 1988 statement for a 20 per cent reduction in CO_2 emissions (HC 192 1989, p.lxx). This is a statement, not an agreement, so it has no legal status. If work is undertaken at a rate of 500,000 low-income homes each year, this will ensure that by 2005 all of the poor are living in homes that they can afford to keep warm and which create less pollution. The annual cost would therefore be £1,250m.

Most of the measures proposed are cost effective, either against the government's 8 per cent RRR (Tables 9.5 and 9.6) or those proposed by Henderson and Shorrock (1990, p.3). And even those measures that are not cost effective, such as some types of solid wall insulation, might become so when the value of additional warmth is included. The costs could be reduced further through bulk contracts or if the work is part of a general rehabilitation. The permutations and combinations possible do not permit more accurate general statements on cost effectiveness. The speed with which the work is undertaken will largely depend on the ability of the manufacturers to supply the materials and the rapidity with which the administrative framework is assembled. Initially, the value of the programme may have to be limited for these reasons, though rapid growth is needed to ensure that all homes are treated by 2005 and to minimize the hardship suffered by those treated last. Some of the work shown here as cost effective may not be undertaken for technical or social reasons: external wall insulation can be architecturally disfiguring and internal wall insulation is very disruptive to the occupants. In these cases, compensatory additional measures will also be needed.

There are several other small programmes already in existence which could be adjusted to supplement the PAW. For instance:

- increased provision of new sheltered housing for elderly people, especially those living in large properties, which are particularly difficult to heat or to convert for a small household;
- the inclusion of insulation measures in enveloping and block repair schemes;
- a greater emphasis on energy efficiency improvements, particularly heating systems, by agencies helping the elderly to 'stay put';
- the minimum level of fitness for a habitable dwelling to be linked to an energy audit level, for instance the 400 MKECI.

An important new component would be the treatment of rented accommodation and this is dealt with in more detail below.

The government's lack of concern and inaction are particularly disturbing as the UK has such a high rate of EWD and low level of energy efficiency, in comparison with other developed countries (Figure 7.6 and Table 10.3). When climate and dwelling size are allowed for, more useful energy is required to achieve a temperature of 14-16°C in the UK than 17-18°C in Denmark. Energy consumption in the UK would be much higher than it is now if we were heating our homes to the temperatures achieved in other countries and do not undertake energy efficiency improvements.

Table 10.3 Useful energy for heating and resultant temperatures, per household, some international comparisons

	Useful energy for heating 1980 (kJ/m²/degree day)	Average indoor winter temperature (24-hour whole house mean, °C)
United Kingdom	230	14-16
Denmark	215	17-18
France	275	17-19
Germany	275	18-20
United States	275	19
Sweden	180	21

Source: Schipper *et al*. 1985, p.3

Involvement

The single most important parameter in the design of a PAW is whether it is to be comprehensive or not: are all low-income households to be treated within a specified time period, or not? Most existing schemes, for instance for loft insulation or draughtproofing, depend upon the individual applying for the grant. If self-identification is used as the basis for the programme, there is every likelihood that some of the most disadvantaged households will fail to apply for assistance, particularly those in the privately rented sector. A policy to assist all fuel poverty sufferers, therefore, has to be comprehensive and include some elements that are universal or mandatory. The importance of energy labels for houses has been emphasized already and is demonstrated in relation to Energy Allowances. Whichever method of energy auditing is used, the possession of an energy label is the simplest way to demonstrate that a property has been included in a PAW. The level of the code indicates how much treatment has been undertaken and how much is still necessary. Thus, the possession of an energy label can be mandatory, without a concomitant requirement to upgrade the house and that is proposed here in relation to the whole housing stock. All other proposals refer solely to low-income households.

A similar procedure has been adopted in Denmark where an energy audit lists cost-effective energy efficiency improvements. When the work has been undertaken and approved, an energy certificate is issued to show that the house has achieved the standard of a new Danish building — equivalent to a 109 MKECI in the three-bedroomed house used in Figure 7.6. Either a certificate or an audit has to be available when the house is sold, and tenants in energy inefficient dwellings are legally entitled to have work done. Between 1981-4, 40 per cent of all homes in Denmark were audited with assistance towards the cost of the audit (*Energy in Europe* 1985, pp.25-8). Thus, energy codes can be introduced rapidly with legislative and financial support. Instead of seeing the Danish scheme as an example to emulate, the British Government's view is that it:

> does not provide an appropriate model for the United Kingdom. It involves an unacceptable element of compulsion and its achievements in Denmark have been largely due to a massive programme of government subsidies for the recommended improvements and formerly for the audits themselves. (Hansard, House of Lords, 2 Nov. 1987, col 780)

A major advantage of a scheme based on specified energy targets is that it incorporates both equity and consistency: it is clear to everyone that priority is being given to the worst houses and the variations in implementation between different areas of the country or local authorities are minimized.

Balancing the different rights of individual fuel poverty sufferers is one of the most difficult tasks for those involved and requires sympathetic and careful handling. Many who are suffering from considerable hardship and fuel poverty may still be fearful of the disruption and possible cost consequences and thus reluctant to agree to work being done. There is no wish to cause distress through enforced disruption or imposed standards, so every effort needs to be made to design the programme so that these natural fears are allayed, the alterations are responsive to individual needs and thus to give the occupant access to the freedom to be warm. This applies to tenants as well as elderly owner occupiers: private and public tenants should be afforded some right of involvement in the decisions affecting the homes they occupy, and for which they will be paying a higher rent.

One way of increasing individual freedom is to permit the work to be undertaken within a specified time period. The example of the Clean Air Act 1956 is that, with sufficient phasing, it is acceptable to require people to undertake alterations to their own homes, and it is proposed that a similar approach underlies a PAW. Where the work can be undertaken at the owner's convenience, the target standard to be achieved can be high. For instance, all homes within an area could be required to achieve a standard of 120 MKECI within, say, five years. Increased flexibility, as with phasing, does limit the cost savings from bulk contracts.

Rented accommodation

Three-quarters of fuel poverty sufferers live in rented accommodation, where many of the problems stem from the conflicting interests of landlords and tenants. Landlords have no legal obligation to improve their premises and additional insulation or replacement heating systems represent improvements. As rent levels are not sensitive enough to reflect variations in energy efficiency, there is no financial incentive for landlords to invest in energy efficiency improvements nor compensation (through lower rents) for tenants if they do not. Similarly,

tenants cannot establish a rental advantage by undertaking energy saving improvements. (ACEC 1983, p.10)

Thus, with the present legal and economic framework, there is no way of encouraging or requiring landlords or tenants to invest in energy efficiency improvements. This largely explains the lack of past investment and the high cost of keeping warm in rented, particularly private, accommodation. The proposal is, therefore, that landlords are made legally responsible for the energy efficiency of their property and any necessary capital expenditure. Therefore, the income of the occupant becomes irrelevant — neither an army General nor a single parent on IS can be expected to invest substantial sums in a rented home. Financial assistance can be made available for low-income landlords in the form of a means-tested grant or equity-sharing loan.

To offset this required capital expenditure, increased rents could be charged. Ensuring that rent levels fairly reflect the energy coding of the property will mean the reinstatement of registered rents and the introduction of a register of rented properties. Without this, it will not be possible to ensure that the rent is fair for both landlord and tenant, before and after improvements.

A potential source of funds for local authority landlords is receipts from council house sales. Since 1983 local authority tenants have had the right to buy their homes and local authorities are permitted to spend only a proportion of the total proceeds from these sales in any one year (25% in

1990/91). The remaining proceeds (about £3bn a year) are retained, on the instructions of the Treasury. The money is available, if the government allows it to be used.

Fuel industries' contribution

If work is undertaken by landlords, this reduces the amount of the required investment needed from central government. The finance to grant-aid work in rented properties and to support other programmes could come from the fuel industries, either as a standard of service requirement or, for electricity only, through the Fossil Fuel Levy (FFL).

Both the gas and electricity industries are required in the privatization legislation to 'promote' the efficient use of their fuels. Effectively this means producing explanatory leaflets, which are of minimal benefit to a poor family with no capital. This requirement could be upgraded if, for instance, both groups of suppliers had to spend a stipulated proportion of their turnover in direct investment in the efficient use of fuel in low-income homes.

An alternative option is to use the FFL being imposed on electricity users. The government has decided that there will be a FFL to subsidize the cost of nuclear-generated electricity. The European Commission has accepted the need for the Levy, provided that the subsidy to nuclear power is phased out by 1998. The Levy will raise an average of £1,150 million pa. The proposal here is that the Levy is retained (with the Commission's permission) and is used to fund an increasing level of energy efficiency investments in low-income households as the proportion that is going to nuclear power diminishes. Thus, by 1998, an annual fund of at least £1,150 million would be available to combat fuel poverty. If this amount was excessive, for instance because of landlord contributions, then the FFL could also be used to encourage the growth of alternative energy.

The use of the FFL has an added environmental advantage. Electricity is the most polluting of the main domestic fuels, as its use, per unit of energy, results in the greatest release of carbon dioxide. The Levy is, effectively, a carbon tax.

Support and administrative framework

It is not appropriate to discuss the detailed implementation of PAW here and only a few more of the main components can be sketched in. The work could be undertaken by a variety of agencies. For instance, it is assumed that the money provided by the fuel industries is retained for investment in low-income households in the same region. The programme of work to be completed in these households could be put out to tender, with local authority direct works departments, community insulation businesses (as promoted by HEES) and others submitting tenders. It is hoped that the fuel suppliers themselves would set up energy efficiency businesses and join in the competitive bidding. This process will ensure that the money derived from the FFL is used to provide the most cost-effective investment in the homes of the fuel poor.

The sequence of dealing with claimants depends on decisions about phasing, savings through bulk orders, self-identification and so forth. One approach would be similar to the North Sea gas conversion programme — working through the country, street by street. The financial status of potential recipients would be confirmed through the receipt of the main passported benefits (IS, HB, community change benefit, family credit — as used for HEES). The aim is that no low-income family should be faced with a legal requirement to do work and a substantial cost. With the proposed PAW, the majority of the costs fall on private landlords or on the local authority. However, where grants are necessary, the proportion of costs covered, or absolute money limit, are crucial in determining the level of take-up. Research in Canada confirms the importance of present costs rather than future savings for low-income households:

> When a household has to worry about the adequacy of their money in terms of covering day-to-day needs, it is not surprising that ... current expenses take precedence over longer-term considerations and results in cashflow dominance. (Claxton *et al.* 1986, p.212)

Where other households receive assistance, for instance enveloping, some of the cost can be clawed back, perhaps through the tax system if this is felt to be necessary.

The co-ordination of this new scheme should be the responsibility of a national agency, of the type first proposed by the Energy Select Committee (HC 402 1982). The agency to be set up to administer HEES would be inappropriate if competitive tendering is to occur. Similarly, the local government Environmental Health Officers could not take on the role, if local authority direct works departments are to tender. An existing possible locus for the scheme could be the regional offices of the EEO.

Another benefit of an extensive energy efficiency programme is its job creation potential. The Prime Minister has argued that increased energy efficiency would be a way, indirectly, of creating more jobs:

> At present we spend £35bn a year on fuel, twice as much as is spent on the National Health Service; £10bn of that is spent by the domestic householder. If only we could cut that amount down — we think by as much as 20% — it would release something like £7bn for purchasing other things. If he has not to spend on energy costs in his house, the householder has money to spend on other things. That in itself could create more jobs. (Excerpt from a speech, Milton Keynes, September 1986, as quoted in ACE 1987, p.7)

There would also be substantial direct employment creation. The best available current source for information on general employment costs in energy conservation work calculated the capital cost of a job created in the European Community as £11-17,000 in 1984 (Hillman and Bollard 1985, p.20). Taking the higher figure, to compensate for the time lapse, a PAW expenditure of £1.25bn p.a. would directly create about 70,000 full-time jobs. These gross costs are substantially reduced at a time of high unemployment by the saving in benefits and generation of taxes. This was estimated to be at least £6,300 p.a. per registered unemployed person in 1984/5 (Sinfield and Fraser 1985, p.12). If these benefits are included and the majority of employees were previously unemployed, the gross cost of the programme is reduced by a third, provided the work created is full time and the employee starts to pay tax. The total cost of any scheme is further reduced by receipts from VAT, which is levied on improvements to existing buildings.

Fuel pricing and the environment

There are several reasons why fuel prices will rise. One of the effects of privatization is to increase the cost of capital and the rate of return that is required. For this single reason, it has been estimated that electricity prices will increase 40-90 per cent over a 25-year period (Bunn and Vlahos 1989, p.114). There will be the same underlying pressure on gas prices.

The pricing formulae that determine the price rises allowed with gas and electricity indicate that electricity prices will rise faster than those of gas: the regulated increases have to be less than the rate of inflation for gas, but can be higher than the RPI for electricity. This will cause additional hardship for the poor because a larger proportion of their expenditure is on electricity and they are more likely to depend on electricity for heating, whether electric CH or on-peak direct acting fires. Other upward price pressures come from world prices of coal or oil, which are notoriously sensitive to political upheaval, and from Europe for consistency of VAT. The UK is the only member of the European Community where domestic fuel is zero-rated (Pearson and Smith 1990, p11). If VAT is imposed, the impact will depend on the rate at which it is levied (15 per cent is the normal rate in the UK), whether it is applied to all fuels, unit costs and standing charges. A full 15 per cent increase on all energy costs would increase the RPI by 1 per cent (*The Times*, 1 Nov. 1984) and have an even greater impact on low-income households, because a higher proportion of their expenditure goes on fuel.

A major problem with all new initiatives will be to ensure that they assist the poor, rather than exacerbate their problems. For instance, referring to the proposals for an internal energy market in Europe, the expectation is that the main benefits will accrue to 'large energy-consuming companies, while the domestic consumer does not appear to benefit either in terms of greater choice or lower prices' (RIIA and SPRU 1989, p.ix).

Any fuel price rise increases the cost of keeping warm and therefore the level of energy efficiency improvements needed in PAW, emphasizing the need for a direct link between benefit rates, fuel prices and the energy efficiency of the home. Any price increase, whether or not the home has been improved, is particularly regressive for low-income families and will often result in decreased consumption and greater hardship. If the cost of fuel goes up, this must be matched by a comparable escalation of investment in measures to improve the efficient use of energy in low-income homes. Otherwise, affordable warmth becomes an even less attainable goal.

The final cause for concern over prices is the potential imposition of a carbon tax. This would be highest on electricity, the greatest polluter of domestic delivered energy (Table 5.3). It has been argued that the money raised from a carbon tax could be used to reduce the impact on low-income households through increased benefits (Pearson and Smith 1990). Because a price rise makes the commodity more expensive, fewer people will purchase it. With fuel, this will be particularly true of low-income households in energy inefficient homes — the cost of warmth is too high to provide sufficient value for the expenditure. Thus, an increase in income equivalent in money terms to the additional cost of the fuel would still be insufficient to compensate low-income households completely. And, it would be going to subsidize the continuing use of fuel in energy inefficient homes.

For the poor, a better and more direct response to the problem of greenhouse gas emissions is direct investment in measures to improve the efficiency with which energy is used. This is a preventative approach, rather than a damage-limitation exercise. The problems of depending on market forces to achieve reduced consumption are addressed by Bowers:

> Getting to (a sustainable growth path) will entail a substantial shift of resources from consumption to investment (in sustainable energy uses, resource conservation and non-polluting production processes). (1990, p.9)

This reinforces the predominant need for direct capital investment, particularly if the poor are not to suffer further.

Maximizing income

Benefits are increased annually in April in line with the RPI in the previous autumn. This is an unsatisfactory measure of the rate of inflation as it affects low-income households, particularly because most fuel increases occur in April and do not, therefore, result in extra benefit for a whole twelve months. A new price index is needed that reflects the purchasing pattern of the poor, including the greater importance of fuel, and that would enable fuel price increases which take effect in April to be included in benefits in the same April.

Paying for fuel and meter choice

A household's success at budgeting is linked to the ability to choose the most appropriate method of payment. It is not certain that consumers do have that right, with either gas or electricity. For instance, in most areas of the country, a prepayment meter is only available for consumers in debt and threatened with disconnection and even then both types of prepayment meter may not be offered. These policies increase the risk of debt and disconnection amongst low-income families and provide further evidence of a failure to consider the needs, and wishes, of poor consumers. The regulators need to protect the consumers' right to choose, so that they are not forced to have credit.

The use of information technology in credit meters is advancing more slowly. Several years ago, Honeywell tested a meter that indicated the cost of energy used since the last bill, but did not proceed with production. The principle of displaying cost information is being tested by some electricity boards with the multi-tariff EMU (energy managment unit). These systems carry up to eight tariffs over the 24-hour period, providing the opportunity for customers to be charged, and to be informed of, a unit price that reflects production costs for that time of day. The period of peak demand is during the winter from 4-7pm. At these times, the price of a kWh is up to 40p (instead of the present 6p). Over the whole year, for an average user (with gas for heating), the total cost of electricity would be the same. However, for all users, the variation in weekly levels of payment would change substantially, causing considerable difficulties for people on a standard benefit. The greatest penalties would be felt by low-income households in all electric homes, so the right of consumers to refuse an EMU, or any other meter, is vital.

Energy allowances

An additional benefit, called here an Energy Allowance, will be needed for at least fifteen years, whilst the PAW is being implemented. The allowance will be needed for longer if during this period there are large, real increases in the cost of fuel or if there are properties that cannot be brought up to the affordable warmth standard. In the meanwhile, the households suffering from fuel poverty are likely to experience greater hardship as a result of both increasing poverty and higher fuel prices. It has to be recognized that there is an interactive relationship between capital expenditure, income and hardship: if there is no increase in the money spent on or available to an individual family, fuel poverty will worsen.

The provision of fuel allowances will be both expensive and environmentally damaging, but is essential given the extent of fuel poverty and the legacy of hard-to-heat housing in Britain. The additional yearly expenditure on benefits and the extra pollution will provide the incentives to keep the capital expenditure programme on target.

At the moment, the social security system does not reflect variations in the cost of keeping warm. Therefore, benefit levels should be related to the energy efficiency of the dwelling and, as fuel prices rise, be increased to reflect the importance of energy costs in the claimant's budget. This is of most importance for low-income owner occupiers, on the assumption that rents are linked to the energy efficiency of the dwelling for tenants.

The £15bn increase in income support is the amount of money required to bring energy expenditure in low-income households up to the level of present fuel expenditure in other households (only £900m actually goes on fuel). At this level of expenditure each of the 6.6m low-income families would be purchasing an additional 20 GJ of useful energy per annum (from Table 8.5). With an average technical efficiency of 68 per cent (from Table 8.3 and Figure 8.2), this would result in an increase in total demand for delivered energy of 200 PJ per annum, or 11 per cent of average domestic energy consumption (Figure 8.1). As this would conflict with the government's stated objective to reduce present expenditure on fuel by 20 per cent — Mrs Thatcher's statement quoted earlier — there would have to be a choice of policy priorities.

The amount of the energy allowance would be determined by the energy code of the dwelling. Two types of energy audit could be used to reflect the different situations in which the Energy Allowance is paid, with the temporary payment based on a simple method of assessment, for economy and speed of implementation. For the permanent payment, however, a more accurate method of assessment may be justified.

In Chapter 3, it is suggested that HB is a more appropriate method of providing assistance to low-income households suffering from fuel poverty than IS and therefore is the benefit to which Energy Allowances are attached. There are several reasons why this is a useful approach, including:

— HB is managed by local authorities, rather than the DSS, which would emphasize the link with housing standards, rather than income support;
— HB is received by householders only, who are, correctly, the main target group for policies based on fuel usage;
— HB is based on actual, as opposed to unquantified, costs, which is similar to the basis needed for assistance with fuel costs;
— HB is tapered which reduces the poverty trap: for every extra £ of income above IS levels, the amount of benefit paid on housing costs is reduced by Xp. This principle could be applied to fuel costs, with only those households with incomes below a certain level receiving 100 per cent of their additional energy costs.

Information and advice

There is one further aspect to helping poor families maximize their income — appropriate information and advice. As identified in Chapter 3, many people, not solely in low-income households, have minimal knowledge about energy use and, often, idiosyncratic beliefs that run counter to a clearer understanding. The provision of good advice, to individual households, would enable these misunderstandings to be cleared up and people to learn how to use energy in their home more wisely. The source should be independent of the main fuel suppliers, to ensure that it is impartial.

Under HEES, network installers (the old community insulation groups) will provide advice to their clients and this is the first time that energy advice has been grant-aided by government. If HEES becomes an established and effective scheme, it provides the opportunity for a considerable extension of energy advice. However, it will never extend beyond clients of the scheme. Two alternative routes are either through Citizens Advice Bureaux, with the possibility of home visits, or through local authorities, probably the housing department. Neither route is ideal: the former might be used mainly by non-low-income households and the latter underutilized by poor households in the private sector. Therefore, some low-income households will be helped with energy advice in future, but the number may only be in the 50-100,000 range each year.

The multiplier effect of expenditure on training advisers could be substantial, particularly if advisers themselves concentrate on training leaders in the community, such as members of tenants associations. This was the approach successfully followed by Optima Energy Services (1989). The need is to improve the accuracy of folk wisdom about new heating technology and appliances, especially through the use of verbal rather than written communication, as:

> nearly all the sample group stated that they preferred to talk about energy use and related issues, rather than *read* about them. (Meyel 1987, p.39)

The wise use of energy depends upon both the availability of information and the ability to understand it. The European Community passed a voluntary Directive on the energy labelling of appliances in 1979, but progress 'has been frustrated by difficulties with establishing satisfactory norms and testing methods' (Jones 1989, p64). The British Government has 'no plans to launch an energy labelling scheme for electrical appliances' (Hansard 29 Feb. 1988, WA col 480). In Britain, therefore, progress has been slow and limited to voluntary argreements and pilot projects, for instance with the Eastern Electricity Board and John Lewis Partnership, despite strong support from the Select Committee on Energy (HC 87 1985, pp.xv, 55). Energy labelling, however, is based on technical measurements and does not include the cost of the fuel. This directive has become incorporated into the Community Action Programme for the efficient use of electricity approved in May 1989 designed to achieve a target of 20 per cent improvement by 1995 (Cm 801 1989, p.24), so more action may occur in Britain.

Conclusions

Fuel poverty can now be redefined as the inability to afford adequate warmth because of the energy inefficiency of the home. Out of the 6.7m poor households in the UK in 1990, 6.6m are suffering from fuel poverty. Without two positive moves by government, fuel poverty in Britain will worsen. First, the government needs to recognize that fuel poverty, capital and revenue expenditure interact at a policy level, just as in the individual family: low levels of investment and fuel price rises result in the need for increased spending on fuel, or greater hardship or both. Secondly, the government has to decide to intervene in the cycle with a level of investment worthy of the problem: billions not millions.

The main proposal is a capital investment Programme for Affordable Warmth (PAW). Because of the low level of energy efficiency of the British housing and heating stock, and because poverty is so extensive, the total investment required is extremely large: £16.5 bn. The money could be raised through different mechanisms, for instance by increasing the responsibility of landlords for the state of their property, as a levy on the fuel industries, or by releasing receipts from council house sales. The cost to government could be minimal, particularly after receipts from VAT on the improvements and the resource savings of employment creation amongst the previously unemployed.

There are additional benefits from the reduction in carbon dioxide emissions, but the real objective is to reduce the suffering of the fuel poor. Evidence of deprivation and hardship are present. All that is needed is the political will to take action, commit sufficient resources and implement an energy efficiency policy to assist low-income families suffering from fuel poverty.

11 Fuel poverty is different

Fuel poverty is different from poverty. General poverty can be reduced through additional income support, but the most effective way to lessen fuel poverty is through capital investment. It is the crucial role of capital stocks — the house, heating system and other energy using equipment — in causing fuel poverty that determines the need for policies that are specific to the problem. A home is energy inefficient, because of a lack of investment and improvement. The occupants, therefore, have to buy expensive warmth and other energy services — they have to pay more to keep warm than people in homes where there has been a higher level of investment in energy efficiency measures. This clear finding shows the fallacy of statements by Ministers that 'there is no such thing as fuel poverty'. Such views only demonstrate the government's lack of understanding of the problem of fuel poverty. An ignorance that is based on a lack of concern and an absence of research.

The old definition of fuel poverty can be extended:

> Fuel poverty is the inability to afford adequate warmth because of the energy inefficiency of the home.

This definition clarifies the importance of capital investment and correctly reduces the role of income support. This is not to ignore the fact that poor families have less money. On the contrary, it takes that as given and recognizes that low-income households should have access to cheap warmth, just as they are able to buy cheap food if they need to in order to exist on their income. This does not mean that present levels of income for the poor are sufficient, only that this is not the place for a discussion on general poverty and its eradication. Depending on the definition of poverty used, up to 99 per cent of the poor are fuel poor.

The perception that the product being purchased is warmth, not fuel, is the key to differentiating between poverty and fuel poverty: the amount of warmth obtained from fuel depends upon the efficiency of two sets of capital stocks — the heating system and the building fabric. Purchasing food or clothing, does not involve capital stocks, but purchasing warmth does.

To provide affordable warmth for an individual household requires the two parameters that are fixed for each household — the income level and the amount of warmth needed — to be balanced by the energy efficiency of the home (Figure 11.1). Just as speed equals distance over time, so does:

warmth = income/cost of warmth

with the cost of warmth measured in £/°C, as in the examples used in Chapter 8. The price of a unit of warmth can only be altered through the energy efficiency of the home, that is changes to COWI factors 1-4.

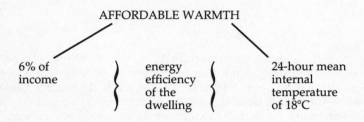

Figure 11.1 Relationship between income, energy efficiency and warmth, if affordable warmth is to be obtained

COWI factors 5-7 summarize the amount of warmth needed by an individual household and this is fixed, like income, if health and comfort are to be obtained. None of the variations in the amount of warmth needed can be influenced by government policy: the hours someone is in the home, his/her age and level of activity and the temperature outside are all independent variables. As most people have similar physiological requirements, a narrow band of temperatures provides comfort conditions for at least 95 per cent of the population. Thus, the quantity of warmth required is only affected by the amount of time spent in the home, not by the desired temperature. The poor are more likely to be in the home and awake for 13 hours a day, because of unemployment, retirement or sickness — nearly twice as long as someone who goes out to work. Thus, the poor have to buy an expensive product to replace the energy services (warmth and lighting) provided free at the work place. This is another distinctive feature of fuel poverty, with few, if any, parallels: there are examples of a lower cost of living for the unemployed, for instance no fares to work, but the rise in energy consumption required by people staying at home is rarely acknowledged.

Another clear finding is that fuel poverty is a problem that affects a large proportion of the population — it is not a minority issue. At least 30 per cent of households are dependent upon the state for their income and have little or no capital. Without money they cannot invest in the energy efficiency of their property and have to depend upon government grants. These 6.7m households are dependent upon the state for both income and capital expenditure.

Why has fuel poverty developed?

Energy efficiency disparities

The role that warmth plays in creating a healthy home environment has never been recognized sufficiently by governments in Britain and this is the major underlying reason for the existence of fuel poverty. Traditionally, our homes have been built with a dismally low level of insulation. Even the improvements in Building Regulations since 1974 are not sufficient to produce energy efficient new homes and there is no systematic programme to ensure that existing homes are improved to a comparable standard. Even when mandatory work is required (as with unfit properties) the level of thermal insulation is ignored.

The other, and more important, aspect of energy efficiency is the calibre of the heating system and the cost of creating warmth in the home. With two relatively minor exceptions, it is not and never has been a requirement of British building or environmental health standards that a domestic building contains a heating system. Near the end of the twentieth century, in one of the most developed countries, there is no government conviction that a home needs to be heated and no concern about the 'affordability' of keeping it warm or the health consequences of cold homes. This appears to be in direct confrontation with *The Universal Declaration of Human Rights 1948*:

Everyone has the right to a standard of living adequate for the health and well being of himself and of his family.

The cumulative impact of variations in the cost of warmth in low-income homes cannot be established in any comprehensive way. A factor of three was identified in the cost of a unit of warmth in one small sub-sample (Figure 7.3) and there is a factor of 20 difference, when standardized for floor area, with the hypothetical examples in Chapter 1. Within low-income homes a ten-fold variation would seem quite plausible.

Low-income homes are less energy efficient and are being improved more slowly than the homes of other families. The reasons include:

— the limited capital and restricted access to cheap credit of poor families prevent them investing in energy efficiency improvements;
— landlords have no incentive to improve and no responsibilities for the energy efficiency of their properties;
— grant-aided government programmes have been restrictive, only covering basic insulation measures. Eligibility was confined to low-income households solely in 1988; prior to this date, most grants were going to better-off families;
— expenditure by government on the two main energy conservation programmes (loft insulation and draughtproofing) is declining in absolute terms;
— in local authority housing, the level of investment has declined since 1980 and is insufficient to offset serious deterioration;

— the Building Regulations are being improved faster than the fitness standard, so that the gap between the best and worst housing in Britain is growing;
— new homes, with better insulation standards, are predominantly occupied by better-off families;
— fuel substitution to heating systems with cheaper running costs has occurred more extensively in better-off homes and has never been grant-aided by government in low-income homes;
— the cost of useful energy varies by a factor of three and the poor are disproportionately likely to use an expensive fuel;
— the poor are more dependent upon electricity than other households and the price of electricity has increased in real terms, whereas gas (the main fuel for other households) is less expensive in real terms in 1990 than it was in 1970.

As a result:

— only 0.5m poor families live in modern (post 1974) dwellings, 2.5m have received some loft insulation or draughtproofing or both, the remaining 3.6m low-income families live in older dwellings with little or no added insulation;
— 59,000 homes in England have no visible method of heating, 1.1m households depend upon general tariff electricity, and a further 0.7m use problematic electric CH, such as underfloor or ceiling heating; 1m use open coal fires, making a total of 2.8m households using expensive systems. 2m have a system that is incapable of heating their home. The overlap between this and the previous group cannot be established. It is assumed that the majority of these are low-income families.

Thus, the differential rate of technical change within the UK housing sector has been to the detriment of the energy efficiency of low-income homes and has been exacerbated by government policies on energy conservation since 1974. Whilst it is certain that, relatively, the fuel poor are worse off in 1990 than they were in 1970, it is also probable that they have suffered an absolute decline in their ability to afford adequate warmth.

Influence of fuel pricing

The impact of fuel prices has been clarified. Fuel price changes have a high profile in fuel poverty debates because of the numbers of people affected and because it is the only way in which the cost of warmth can be substantially increased. Other factors that increase the energy inefficiency of a dwelling are changing at an imperceptibly slow rate (such as building fabric deterioration) and may not be affecting all the poor. The short-term effect of a fuel price increase is to compound budgeting difficulties: a sudden rise causes particular hardship and dismay. For those on low incomes higher fuel prices reduce disposable income and hence the ability to respond through investment in energy efficiency improvements. In addition, in energy inefficient homes the cost of warmth is already high, so any increase in fuel prices means that a larger absolute amount of money has to be found if temperatures are to be maintained. Thus, in reality, real fuel price increases do cause greater hardship for the poor than other families, because of their energy inefficient homes and their inability to respond (for financial and legal reasons).

The pricing formulae of nationalized and privatized supply industries appear likely to perpetuate these relativities and thus large, and perhaps growing, fuel price differences will remain an important component of variations in energy efficiency.

Great care is needed to distinguish between the demand for warmth and the demand for energy, as these can be of quite different magnitudes — and even opposite directions — in the same household, at the same time. Evidence from income elasticities indicates that most households, but especially the poor, will use additional income to purchase more fuel. This confirms that there is a general desire for greater warmth. However, the economically rational response for someone on a low income when faced with an expensive commodity is to purchase little of it. This is exactly what many poor families in energy inefficient homes do.

There is futher evidence of confusion over the economic benefits of investing in energy efficiency measures in cold homes. The policies followed by the DEn and EEO do not recognize that the increased warmth experienced by the occupants after improvements represents a real resource benefit, in economic analyses. This is contrary to the principles followed by both the DOE and the Ministry of Defence, where additional warmth is an acknowledged benefit. Thus, investment seen as complying with government economic criteria by the DOE or MOD would be rejected by the DEn.

Increasing poverty

The existing situation can be summarized by evidence of government expenditure: the total spent on capital improvements to the energy efficiency of low-income households over 13 years is £300m. Annual expenditure on fuel by these same households out of state-derived benefits is £2.6bn. The latter increases by the rate of inflation, indefinitely, whereas capital expenditure is decreasing. This level of revenue expenditure by the government is double, in real terms, the level in 1970, mainly because of the increasing numbers of people dependent upon the state. The majority of the poor are in receipt of HB and are having their weekly housing costs supported, as well. Thus, the state is helping them to pay for space they cannot use, because it is cold, due to a lack of further resources to make the home energy efficient.

The increase in fuel poverty is closed linked with the growth of poverty. The evidence includes:

— the real incomes of the poorest 20 per cent declined by 1.1 per cent between 1979-87, whereas for the rest of the population they grew by 29 per cent;

— there were about 3.9m families dependent upon the state in 1973, whereas the number is estimated to be 6.7m in 1990;

— income-related fuel poverty has grown faster than the numbers in poverty: there were 300,000 households with the most acute fuel bill problems in 1977, but 540,000 in 1989;

— the number of households paying off a gas debt doubled between 1985 and 1990 to 2m;

— over 2.5m households threatened with disconnection have had a prepayment meter installed between 1977-89, reducing the numbers still at risk of fuel debt;

— benefit levels do not vary according to any aspect of energy efficiency, not even to reflect standard temperature variations in the different parts of the country;

— the abolition of heating additions in April 1988 resulted in the poorest families losing at least £200m. This was money that had previously been paid to compensate for the energy inefficiency of the home particularly related to heating systems and fuel;

— the only remaining weekly benefit associated with fuel costs is being phased out and now goes to a few thousand households paying for heating in their rent;

— assistance with fuel consumption is specifically not allowed from the Social Fund; the exceptionally severe weather payment has to be claimed and carries a 50 per cent administrative cost.

As a result:

— the number of excess winter deaths in the UK is still in the 30,000-60,000 per winter; there are an extra 8,000 deaths for every °C that the winter is colder than average;

— temperatures in British homes are well below those of our European neighbours, but take more energy to attain;

— by 1988, low-income households were able to purchase at least 22 per cent less warmth than they needed in comparison with better-off families. The figure was 18 per cent in 1985, indicating the widening disparities.

The uncertainty about the actual cost of warmth demonstrates the lack of attention given to the subject in the past. The only other necessity to demonstrate a comparable range is housing, where the cost for low-income households is covered by up to 100 per cent subsidy and increased expenditure is closely related to better quality housing. According to the Treasury, it is because housing costs vary so much, even for similar properties, that they have been isolated from income support. Assistance with fuel purchases is minimal and is being phased out, and expensive warmth does not confer greater utility on the purchaser than cheap warmth — it represents a misuse of resources for the individual and the nation. Every household in the country obtains 1.25 MJ of energy from a

24p tin of baked beans. If this were true of expenditure on heating as well, solutions to fuel poverty would be the same as those for general poverty — additional income.

Because the poor live in energy inefficient homes they are having to buy the most expensive warmth. This is the reverse of normal expectations that the poor cope with a shortage of money by buying inexpensive goods: shopping around for special offers or cheaper cuts of meat. Thus the poor, with less money, have homes that are more expensive to heat, and are unable to use a cheaper fuel or otherwise obtain better value for the money spent on heating without capital expenditure. It is because the opportunities for substitution are so severely restricted that warmth is such an unusual commodity. In the short term, each household is only able to buy warmth at a single price — shopping around is only possible with coal — and for the poor in energy inefficient homes the price of warmth is high. Because they are in the home so much, low-income households need more warmth, but each unit of warmth costs them more, and they have less money with which to buy it. That is the kernel of the fuel poverty problem.

Solutions

A Programme for Affordable Warmth (PAW) is needed to bring all homes occupied by a low-income household up to a standard that allows them to have adequate energy services for 10 per cent of income. This appears to be a 120 MKECI, NHER of 8 or similar and means that all the 6.6m homes of the fuel poor should be improved to the standard of the 1990 Building Regulations, together with double glazing and gas CH, or equivalent. The total cost of this programme would be £16,500m, to be completed by 2005. The poorest 30 per cent are responsible for only 24 per cent of domestic carbon dioxide emissions and this programme would reduce the present level of 6.4 tonnes of CO_2 per household to less than 6 tonnes per annum.

It is possible to reduce fuel poverty through additional income support, but this would cost about £15bn each year, increasing in line with inflation. The resultant extra energy demand from low-income households would be equivalent to an increase in total domestic delivered energy consumption of about 11 per cent, with the concomitant need for extra growth in capital expenditure in energy supply. This scenario is viewed as environmentally and economically unrealistic.

The energy efficiency of the dwelling has to be defined by an energy audit using a defined heating standard and appropriate levels of other energy services. An energy label gives a clear and non-discriminatory basis for policy and allows the worst housing to be identified and improved first. It also identifies that a building has been seen and incorporated into any programmes and provides the basis for additional income, an Energy Allowance, for low-income families before their property is treated.

The PAW would involve the upgrading of rented accommodation through regulations imposed on landlords, public and private. The

standard of an unfit home would be progressively increased from a
minimum energy code of, say, a 400 MKECI. Other supportive
programmes include the provision of more sheltered housing and a
greater emphasis on heating system improvements in all rehabilitation
schemes. Several other policies have been identified that could be
implemented quickly to provide assistance to fuel poverty sufferers,
including:

— upgrading the estates previously designated as expensive to heat
  (ERHA);
— extending HEES to include cavity wall insulation grants for low-
  income households and making sure that it has an adequate
  budget, at least two or three times the £12m for six months original
  budget;
— introducing a grant to ensure that there is a fixed heating system
  within low-income households that is capable of heating the
  dwelling at reasonable cost;
— providing fuel substitution grants for households dependent
  upon general tariff electricity for heating, so that they can obtain a
  less expensive and less polluting fuel or tariff;
— each Regulator to undertake an assessment of the apportionment
  of costs between the standing charges and unit rates for both
  electricity and gas to establish if tariff tilting would benefit the
  poor and to clarify the justification for high charges on token
  prepayment meters;
— enabling all households to obtain their gas or electricity supply
  through the use of a prepayment meter, if they want — they
  should not be forced to have credit;
— varying the level of income support in 10 per cent bands to reflect
  the additional weather-related energy costs of different regions, as
  proposed by Gordon Wilson MP;
— increasing benefit rates in line with a new price index that reflects
  the expenditure patterns of low-income households more
  accurately.

Because the PAW will take several years to complete, there is an interim
need for additional income support, or energy allowance. Provided that
this is related to the energy efficiency of the dwelling, it could be paid on
the basis of a simple energy audit and not actual expenditure. Additional
income support is a necessary palliative, though immensely expensive
and would result in increased energy consumption without any
improvement in energy efficiency. It has the added disadvantage of
increasing the emission of greenhouse gases thus providing the greatest
imperative for the rapid implementation of PAW.

Unfortunately, the problem of fuel poverty is not going to disappear.
During a mild winter, the problem will stay submerged, despite
deteriorating conditions for the poor. But the increase in winter deaths
and hospital admissions that will occur during a spell of cold weather as
severe as January 1987 or February 1986 will be met by short-term
responses, with no lasting impact on the causes of fuel poverty.

There are some rays of hope. The government's emphasis on value for money has resulted in the more effective targeting of grants (for loft insulation, home improvements and draughtproofing) on the poor, though there is uncertainty about the adequacy of maximum payments or total allocations. Energy efficiency is, gradually, becoming an integral part of housing standards, hence a fixed heating system in the main living room is a component of the fitness standard. A further imperative is coming from European Community pressures for domestic energy efficiency.

The government's fragmented approach towards energy efficiency, and the resultant split of concerns between government departments, is seriously inhibiting the development of appropriate policies for low-income households. The first step toward progress on combating fuel poverty is for the lead department to be clearly identified. The necessary emphasis on housing quality, the links with HB, the role of local authorities as administrators and the pivotal importance of landlord and tenant law, all point to the DOE as an appropriate co-ordinator. The DEN has, however, taken responsibility for the HEES.

The present, deteriorating situation demonstrates another aspect of the interrelationship of factors affecting fuel poverty. If there is no co-ordination, there is the likelihood that benefits in one area will be annulled by increased hardship in another. Where it is accepted that a change is inevitable for other, wider policy reasons, for instance an electricity price increase, then compensatory energy efficiency policies are necessary. The difficulties inherent in providing adequate compensatory policies have to be recognized. Real fuel prices have risen 12 per cent since 1970, for all 6.7m low-income families, whereas the work of community insulation groups has achieved a 5 per cent reduction in the fuel bills of about 750,000 low-income homes as a result of 13 years' work. A unilateral fuel price rise, affecting all users, can only be offset by many years of hard work in millions of homes.

The future of the fuel poor will depend upon a range of political choices. The evidence indicates that fuel poverty is increasing, as a result both of minimal intervention and, by default, because of the continuing impact of recent decisions. The present situation is not static, so inaction by government is equivalent to an acceptance of a worsening situation.

In practice, government policy will probably result from the interaction of several criteria: restraint on government expenditure; concern about energy imports and the effect on the balance of payments; job creation; European Community initiatives; or, the need to reduce energy consumption because of global warming. The drive of nearly all of these will be to encourage investment in energy efficiency measures at nil expense to government. In the short period before the electricity industry flotation is completed, there will not be a great deal of discussion about the potential for energy efficiency, as this would undermine the value of the shares. After April 1991, the government will be able to discuss demand reduction more freely.

The more efficient use of energy is an important component of all solutions to global warming and environmental concerns may be the most likely reason for new policy initiatives on energy efficiency. However, any response that is predicated upon higher fuel prices, for instance as a result of carbon taxes, will increase the relative deprivation of the poor. Policies to reduce fuel poverty and to lessen the amount of carbon dioxide emitted depend upon capital investment. No other response achieves both objectives.

Fuel poverty is a serious social problem that continues despite considerable state expenditure. Policies have been inadequate and misdirected, partly because they were based on an analysis of the symptoms of fuel poverty, rather than the causes. With social problems needing inter-disciplinary study, the events that trigger public concern, such as the numbers of disconnections or deaths from hypothermia, can be poor indicators of the true causes and necessary solutions. The energy efficiency of the home environment is what determines whether a low-income family can obtain adequate warmth. It is because the poor in Britain have no choice but to purchase expensive warmth that they suffer from fuel poverty.

Glossary and notes

BEDROOM STANDARD — According to Gray and Russell (1962, p75) this is calculated as follows:

i. each married couple is given one bedroom;
ii. any other persons aged 21 or over are given one bedroom each;
iii. unmarried persons aged 10-20 of the same sex are given one bedroom per pair;
iv. any remaining person aged 10-20 is given a bedroom with a child aged 0-9 of the same sex; if this is not possible he or she is given a separate bedroom;
v. any remaining children aged 0-9 are given one bedroom per pair; any unpaired child is given a separate bedroom.

BREDEM — Building Research Establishment's Domestic Energy Model:

is a model for the calculation of the annual energy requirements of domestic buildings, and for the estimation of savings resulting from energy conservation measures. (Anderson *et al.* 1985, p.vi)

Several versions of BREDEM have been developed, particularly by Energy Advisory Services, for instance Energy Auditor is for existing buildings and Energy Designer for new build.

BREHOMES — A computer model of the heat loss characteristics of the housing stock, assembled for the Building Research Establishment, and including variations with time. The main groups are dwelling type, age, tenure and the presence or absence of CH. There are no income variables.

BUDGET WARMTH — This scheme is offered by some PES, principally to elderly local authority tenants. In return for a fixed weekly payment, one (or more) night storage heater is installed in the dwelling and the amount of charge controlled via radio signals, by the local board, to match the weather conditions. Although no insulation is installed, the expectation is that the occupant has at least one warm room. Because there is no metering, the average cost of fuel used is difficult to assess, but could be high. The night storage heaters remain the property of the Board. As they have to pay for consumption, it is hoped that this scheme will prevent elderly people economizing on heating to a dangerous extent.

BUILDING METHODS AND MATERIALS — The following

definitions have been used, based on *Roof*, November/December 1986, p.15:

Traditional building — Brick built on site;

Non-traditional building — New materials such as plastic, metal and concrete used, generally constructed on site;

Industrialized buildings — The building components are prefabricated or constructed at the site using mechanical plant. Both the design of the building and construction process are, therefore, non-traditional (AMA 1983, p.3);

System building — The whole design and construction process is purchased as a package; considerable amounts of the building are usually industrialized. Most system built homes are named after the firm who built them, e.g. Wimpeys, Boots, Orlit or Bisons;

Tower blocks — Any dwelling over four storeys, however constructed.

Because many system built homes are causing major problems for tenants, the convention has grown of describing most non-traditional estates with serious defects as 'system built'. Thus, the phrase is now indicative of tenant dissatisfaction as much as a true description of the building method.

CAVITY WALL INSULATION — There are several methods of filling an existing cavity, including foam, beads and fibre. The latter has the widest performance range (including exposed buildings) and is not thought to affect the health of the occupants, though it is the most expensive. Through bulk contracts, costs can be reduced from £7/m² to £2/m²

CENTRAL HEATING — The general principle is that heating is provided from one central source, usually within the dwelling, to one or more outlets. Thus, a single night storage heater is defined, in most statistical sources, as central heating. There is virtually no data on the number of outlets provided by CH systems, either by type or over time. (For district heating see combined heat and power below.)

CLO — The insulation provided by clothing is measured in 'clos', with 1 clo equal to a resistance of 0.155 m² °C/W (Humphreys 1976, p.177) — the inverse of U values for building components, so the larger the number of clos, the greater the level of insulation provided. Normal winter wear of skirt and jumper, trousers and jumper, three-piece suit or warm dress provides from 0.8-1.2 clo (BRE 1979, p.2).

CODE OF PRACTICE — the electricity and gas industries agreed to a voluntary Code of Practice, effective from January 1977, to minimize hardship and clarify consumers' rights. This has been partially replaced by conditions in the supply licences, so that consumer protection is now a mixture of a voluntary code and statutory protection. The industries are now producing separate codes.

COMBINED HEAT AND POWER (CHP) AND DISTRICT HEATING — With CHP, the hot water produced in the generation of electricity is used to heat buildings through a district heating system. Most discussion is about CHP in relation to existing or proposed power stations, but smaller schemes are possible. With district heating there is no electricity generation. The cost of heating is either paid for by a standard charge or is metered per dwelling. Much of the dissatisfaction expressed by tenants with district heating relates to the apportionment of costs.

COWI — The Cost of Warmth Index is calculated in £ per day:

$(1 + 2) \div 3 \times 4 \times (6 - 5) \times 7$ where
1 is the rate of heat loss through the building fabric (kW/°C)
2 is the rate of ventilation loss (kW/°C)
3 is the efficiency of the heating system (%)
4 is the cost of the fuel used (£/kWh)
5 is the external temperature, adjusted for incidental gains (°C)
6 is the internal temperature required (°C)
7 is the hours of heating per day (hours)

Thus, factors 1-4 express the energy efficiency of the dwelling (the cost per unit of warmth) and factors 5-7 quantify the demand for warmth. Other units of measurement can be used, with suitable adjustment, for instance gigajoules, degree days and room-hours of warmth.

DEGREE DAYS — To measure degree days, the temperature fluctuations over a 24-hour period are related to a base figure: each degree day represents 24 hours x 1°C below this base. In the UK the base figure is generally taken as 15.5°C, because, based on research in offices, it is estimated that incidental gains are equivalent to a 2.8°C increase, bringing the 24-hour MIT to 18.3°C. (The original measurements were in Fahrenheit, giving a MIT of 68°F.) Historically, when degree days were being defined, a MIT of 18.3°C was assumed to provide comfort conditions. The base temperature appears to overestimate the amount of incidental gains likely in the average home, particularly a low-income one, so that a higher base should be used when calculating comfort conditions in a British home.

DELIVERED ENERGY — Energy in the form that it is delivered to the house, before it is converted into useful energy in an appliance.

DOUBLE GLAZING — This usually means that the original window is replaced with a sealed unit containing two panes of glass. Installers are usually required and the expense is considerable and only cost-effective if the window needed replacing anyway. Cheaper processes, involving the retention of the existing window, are referred to as secondary glazing. Secondary glazing has its own frame set within the window recess behind an existing window; this can be DIY.

DWELLING — Means the space occupied by a household and thus can include a flat. In the UK the total number of dwellings exceeds the number of households, implying a surplus. However, some dwellings are second homes, vacant or undergoing conversion so the inhabited total is substantially less, though there is no firm estimate of this figure (*Social Trends* 1986, p.134), and thus it is not clear whether, over the whole country, there is a shortage or surplus in relation to the numbers of households. Certainly, in some regions, such as London, there is an acute shortage of publicly rented accommodation available for low-income households, particularly when they become homeless. For these reasons, most data in the book relate to households of people, rather than dwellings.

DWELLING SIZE — There are no comprehensive national statistics on the size of the dwellings in the housing stock. The average, older dwelling is generally larger than more modern buildings (Leach *et al.* 1979, p.97; Evans & Herring 1989), partly because of the smaller number of flats built earlier. However, the floor area of ostensibly identical buildings has fluctuated considerably. For instance, local authority homes with three bedrooms had an average area of $70m^2$ in the 1930s, which rose by 40 per cent to $98m^2$ between 1945-51 (Merrett 1979, p.322). The average size of new local authority buildings is again declining (Housing and Construction Statistics). Floor areas influence the amount of external fabric, and therefore, heat loss, as well as the volume and ventilation loss.

ELECTRICITY — There are 12 Public Electricity Suppliers (PES) in England and Wales, which are subject to one form of privatization. The South of Scotland Electricity Board, North of Scotland Hydro-Electric Board and the Northern Ireland Electricity Board are being treated slightly differently.

ELECTRICITY TARIFFS — The general or standard tariff applies throughout the 24-hour period with the ordinary meter, and is known as the unrestricted rate. The less expensive off-peak or restricted tariff is available only during certain hours at night, with the addition of another meter and higher standing charge. The present off-peak rate is known as Economy 7 in England and Wales and White Meter in Scotland, and is available for 7 hours of the Board's choosing, but roughly midnight to 7 am. Previous off-peak rates, that are no longer available, are known as historic or preserved off-peak rates; although the off-peak rates are higher than for Economy 7, they often include an afternoon 'boost' at a concessionary rate.

EQUIVALENT INCOME — The European Commission uses disposable income to measure poverty, but with adjustments for family composition. The members of the family are weighted according to the scale used in Britain, for instance with the old Supplementary Benefit (Cmnd 9519 1985, p.17):

married couple without children = 100;
single householder = 61;
married couple with 2 children (aged 4 and 6) = 139.

This 'equivalent income' gives a better indication of the standard of living that can be achieved by the household, than the unadjusted disposable income (Berthoud *et al.*, 1980, p.21), partly because it removes the bias towards small households. However, equivalent income data are normally computed specially and are not widely published.

ESTATE RATE HEATING ADDITION (ERHA) — Estates of houses with heating systems that are disproportionately expensive to run could be designated, so that residents who were SB claimants could receive this HA. The majority of qualifying heating systems use electricity on a preserved (or historic) tariff. This was the only HA or social security benefit to be related to the cost of the fuel used, and it was abolished in April 1988.

EXCEPTIONALLY SEVERE WEATHER (ESW) — This phrase has a precise meaning for the DSS, as it is used to trigger additional payments to IS claimants. The definition has, however, varied almost every year, but for the present scheme a period of ESW exists when the average temperature over seven consecutive days is below 0°C. The payment is £5 per week for households containing anyone who is over 60, under 5 years or chronically sick or disabled. There is a capital cut-off of £1000. This benefit always has to be claimed, retrospectively.

EXCESS WINTER DEATHS (EWD) — There are several different methods of measuring excess winter deaths and no way of making comparisons between them without access to the original data. The main methods are based on one, three, four or six monthly periods:

monthly — the number of deaths recorded per month are calculated as a proportion of the number that would have been expected, if they were equally distributed throughout the year. The twelve resultant percentages are used to compute a coefficient of variation, as in Table 7.2 and figure 7.5. This coefficient is used for international comparisons, for instance in United Nations publications;

three monthly — the number of deaths in the winter quarter (January-March) are compared with the number in other quarters, or in the summer quarter only (July-September). The most easily obtained from published OPCS statistics, for instance in the Monitor;

four monthly — the number of deaths in the four winter months (December-March) is compared with the number of deaths in four non-winter months. The latter is obtained by averaging the figures for the preceding August-November and the following April-July periods. This comparison is either given as an actual number of excess winter deaths (as in the equation devised by Curwen 1981, quoted in Chapter 7), or is shown as a percentage increase;

six monthly — the winter period of October-March is given as a percentage of the summer (April-September). Not widely used, but has been quoted in Hansard.

FAMILY — This includes one-person households.

FAMILY EXPENDITURE SURVEY (FES) AND FUEL COSTS — Because of the survey method, the data on gas and electricity expenditure is fairly reliable, whereas information on purchases of solid fuel, paraffin and wood is less robust. No data are published on the breakdown in electricity expenditure between on and off-peak tariffs. A detailed analysis has been undertaken by the Social Policy Research Unit (1981).

FAMILY EXPENDITURE SURVEY (FES) AND HOUSING COSTS — The way the FES treats housing expenditure creates a serious limitation on the usefulness of the data and has even caused the government confusion. Owner occupiers' expenditure on mortgages is not shown in the FES because 'part of their payments can be regarded as savings rather than current expenditure' (FES 1984, p.95). Instead, an imputed rent, based on the rateable value, is used for housing expenditure, though this is lower than mortgage payments for many households. Thus the FES gives an artificially low level of housing expenditure for households with mortgages, and a parallel reduction in total spending. Equivalent procedures are followed for people in rent-free accommodation and full owner occupiers, where the imputed rent results in both housing expenditure and total spending appearing to be higher than they are in reality. Most low-income families are paying for rented accommodation where the FES shows actual rent, rather than some imputed figure. There are two complications for these low-income families. The first relates to housing costs as reflected in the RPI (see below) and the second to recent changes in the reporting of rebates and allowances for housing costs.

The problem with the reporting of housing benefits has occurred in Reports since 1982. Under both the present Housing Benefit (HB) scheme and the previous system of rent and rate rebates and allowances, households eligible for housing assistance either receive it as extra income or have the rebate applied at source through lower rent and rate demands, as the money is paid direct to the landlord. Thus, these similar housing subsidies can result in additional income or in lower expenditure. In 1982 and earlier Reports, the FES, where possible, showed all housing benefits as increases in income, with housing costs at the gross level. Since the introduction of HB, the FES has revised its procedures and now reduces housing expenditure to the amount paid net of benefit, and leaves income at the original level. This is the procedure followed in the 1984 and subsequent FES Reports. Unfortunately, there was a further complication with the 1983 FES. HB had a phased introduction (from November 1982) and the administration is now handled entirely by local authorities. Considerable delays in handling cases occurred, so many recipients did not receive their HB, particularly in the first half of the year. As a result, reported housing expenditure by low-income households was confused and the 1983 results are suspect for this reason (DOE pers comm). The problems created by these changes are clear from the following figures for housing and all items for the lowest income quintile:

1982	£13.83 out of £59.21 is 23.4 per cent on housing
1983	£13.24 out of £58.45 is 22.9 per cent on housing
1984	£ 7.59 out of £53.44 is 14.2 per cent on housing
1985	£ 7.78 out of £56.51 is 13.8 per cent on housing

Because of the way that the FES treats housing expenditure, the total expenditure shown is a hypothetical figure, for all households in owner occupation or in rent-free accommodation. Thus, the amount spent on other items, such as fuel or food, are only given as a proportion of a hypothetical total expenditure. For low-income households, the change to net housing expenditure in 1984 means that comparisons of expenditure on items such as fuel, over this period, must be treated with great caution.

FITNESS STANDARD — see Unfit for human habitation.

FUEL DIRECT — IS claimants, with the agreement of the DSS and the appropriate fuel industry, can choose to repay a fuel debt through direct deductions, at source, from their benefit entitlement. This process is known as being on fuel direct. The money deducted each week includes a maximum amount for the debt repayment (currently £3.70) together with the sum necessary to cover current expenditure. If there are debts with both gas and electricity, the repayment rate is halved (Lorber *et al*. 1989, p.76).

FUEL PRICE INDEX (FPI) — The cost of the major domestic fuels (excluding petrol) are weighted by current use to give the fuel price index, a separate component of the retail price index (see RPI below).

GAS — Where data are obtained from British Gas this refers to Great Britain. However, there is no natural gas in Northern Ireland and limited use of town gas, so that British Gas statistics are, effectively, covering the UK. Some statistics include consumption of calor gas with natural and town gas, but relatively few households are affected.

HEATING ADDITIONS (HA) — Assistance with heating costs was given as part of National Assistance at least from 1965 onwards. Under the replacement, SB system, payments were either weekly additions (exceptional circumstances addition or ECA) or one-off, exceptional needs payments (ENPs). Originally, both were given on a discretionary basis, then revisions in 1980 resulted in a legal entitlement that was defined in regulations. Exceptional needs payments were substantially phased out when the voluntary Code of Practice was introduced in 1977, though there is a residual need, covered now by grants from the Social Fund for reconnection costs and moving meters. HA payments ceased in April 1988, when the Social Security Act 1986 was implemented.

HOUSEHOLDS AND INCOME UNITS — In this study, the occupant or occupants of a dwelling can be called either a household or a family — the terms are interchangeable. Within a family, there can be several people who are individually entitled to IS; for instance, a couple and their non-dependent child are two income units. For this reason, there are more IS 'income units' than households and many IS claimants are non-householders (Figure 3.3). No work has been done on income units here.

The number of households in the UK is not a published figure and depends upon projections from census information The numbers used throughout this study are given in Table A.1.

HOUSING BENEFIT (HB) — Housing benefit is means-tested, but is separate from the main social security system, as it is administered by local authorities. HB was introduced in April 1983 and replaced the previous systems of rent and rate allowances and rebates. Until April 1988, support for rent and rates were handled separately, whereas now they are combined. HB is assessed in relation to actual costs and above a specified income the amount of HB is tapered, by a standard amount for each extra £ of income. Most reductions in numbers of HB recepients have come from increasing the proportion in the taper.

Table A.1 Private households, UK 1970-90 (millions)

1970	18.7	1971	19.0	1972	19.2
1973	19.3	1974	19.4	1975	19.6
1976	19.7	1977	19.8	1978	20.0
1979	20.1	1980	20.4	1981	20.6
1982	20.7	1983	20.9	1984	21.1
1985	21.3	1986	21.5	1987	21.7
1988	21.9	1989	22.1	1990	22.3

Source: DOE, pers comm

HOUSING INVESTMENT PROGRAMME — Each year local authorities draw up a housing expenditure budget, based on their Housing Investment Programme. Before 1980, this included a specific budget head for energy conservation, so that money for the local authority energy conservation programme was specifically, and solely, for that purpose. It was hypothecated, or 'earmarked'.

INCOME SUPPORT — IS represents the minimum income level, below which no household is supposed to fall and replaced supplementary benefit (SB) in April 1988.

JOULE — Most energy reports now use the SI unit of a Joule (1 Watt per second). $1kWh = 3.6 \, 10^6 J$.

MINOR WORKS GRANT — A component of the new Housing Renewal Grants which replaced Home Improvement Grants on April 1990. It is for £1,000 and can be claimed up to three times. Claims to improve insulation are specifically encouraged, though improvements to heating systems can only be funded for elderly people. Local authorities have discretion as to whether they make money available for the MWG, though it replaces the loft insulation scheme, which was a universal grant, statutorily available. Claimants of the main means-tested benefits (IS, HB, Community Charge Benefit and Family Credit) are eligible to claim.

MKECI — In the Milton Keynes Energy Cost Index, the total energy expenditure (including standing charges and maintenance) required to obtain a specified standard is related to the floor area of the house and converted into an index. Houses with the same level of energy efficiency should achieve the same coding, despite changes in fuel costs. Low numbers are more energy efficient.

NHER — National Home Energy Rating, like MKECI, measures the energy efficiency of a home. NHER uses a scale of 0-10 with 10 being the most energy efficient.

NET CALORIFIC VALUE — The calorific value is the amount of heat evolved when unit mass of a fuel is burned completely under standard conditions. Gross calorific value (GCV) is the amount of heat evolved in a bomb calorimeter when the products of combustion are cooled and water vapour has condensed, as in a condensing boiler. With the lower net calorific value (NCV) the latent heat of condensing water is deducted from the GCV and thus reflects more accurately the heat evolved under most operating conditions (Slesser 1982, p.36).

NO FINES — A form of concrete containing no fine sand and mainly consisting of small stones cemented together (giving it a similar appearance to chocolate covered rice krispies). If no fines concrete is correctly mixed and installed on site, it is supposed to have good insulation qualities because of the large proportion of the volume that is air. There is some controversy as to whether the material performs in this way in practice.

NORTHERN IRELAND — Every attempt has been made to include information on Northern Ireland wherever possible, but most policies and statistics concerning the province are dealt with by the Northern Ireland Office, rather than Departments concerned with the mainland or England and Wales only. Collecting comprehensive statistics for the UK is, therefore, a daunting task. Fuel poverty is particularly bad in Northern Ireland, because of the absence of natural gas and the relatively high price of electricity (under existing government policy, the price cannot be higher than the most expensive region in England and Wales).

NOTIONAL FUEL ELEMENT (NFE) — The DSS do not identify how much money is for fuel within the scale rates, or other benefits. However, in 1987 about 90,000 SB claimants paid rent inclusive of some or all fuel charges. Therefore, for these households to receive HB, the amount of money in their benefits for these fuel costs has to be identified, so that they are not receiving assistance with the same expenditure through both HB and SB/IS. The NFE is thus the unofficial statement of how much money was in SB for fuel, and it is the same fixed amount for all groups and types of SB beneficiary, regardless of fuel used or housing circumstances. Within the total of £8.80 (July 1987), there were the following elements:

Heating	£6.70 (76%)
Hot water	£0.80 (9%)
Cooking	£0.80 (9%)
Lighting	£0.50 (6%)

These amounts are unrelated to real expenditure by the claimant family. Even this unofficial figure is to disappear from benefit vocabulary as it is being phased out from April 1988 onwards, because it represents unfair discrimination in favour of SB claimants. The average weekly payment of £2.50 is being reduced at a rate of £1.50 a year, so by 1990 very few families receive 'topping up' above NFE. The amount of NFE does not, therefore, need to be declared, so it will cease as a method of unofficial guidance on average fuel costs.

OFFER — The Office of Electricity Regulation, headed by the Director General of Electricity Supply, is responsible for regulating the whole of the electricity industry since privatization. The DG has a major responsibility for consumer protection.

OFGAS — the Office of Gas Supply, headed by the Director General, is responsible for regulation the gas industry since privatization. The Gas Consumers' Council provides assistance with consumer affairs, but has no statutory powers.

OPCS (Office of Population Censuses and Surveys) — This is the major source of mortality statistics, but only covers England and Wales. Where possible, data for Scotland and Northern Ireland have been obtained on seasonal deaths, but the coverage is not complete. Therefore, care is needed when comparing statistics, because of the different areas involved.

PAYBACK — The payback period (in years) on an investment is obtained by dividing the benefits into the cost. This is the simplest form of return and ignores the cost of borrowing money or possible price rises.

PENSIONERS AND THE VERY ELDERLY — A woman becomes eligible for the state pension at age 60 and a man at 65, and everyone over these ages is termed a pensioner. Some DSS policies (for instance ESW payments) are for people over 60, thus including non-pensioner men aged 60-64. The very elderly, in this thesis, are those pensioners, of both sexes, aged over 75, though in some other contexts the age of 85 is used.

PES — The old area electricity boards are called Public Electricity Suppliers (PES), now that they have been privatized. They can also be called Discos (Distribution Companies) or RECS (Regional Electricity Companies).

POOR — In this book the poor are defined as those households who depend upon the state for over 75 per cent of their income — a virtually identical group to the 30 per cent of households with the lowest incomes (see Chapter 3).

PRIVATELY RENTED — The subdivisions within the privately rented sector (furnished and unfurnished) represent two very different groups. For instance, the income of the upper quartile in furnished privately rented accommodation is more than three times higher than that of the lower quartile, a greater spread than in any other tenure (GHS 1984, p.55). The age of tenants in furnished accommodation is much younger than those in the unfurnished sector. More comparisons are available from the GHS.

PUBLIC SECTOR HOUSING — This is assumed to include housing provided by housing associations and co-operatives, as well as local authorities, unless stated to the contrary.

REGULATOR — see Offer and Ofgas above.

RETAIL PRICE INDEX (RPI) — The General Index of Retail Prices measures the monthly changes in the prices of commodities and services purchased by all types of households in the UK, with two exceptions. The expenditure pattern of the 3-4 per cent of households with the highest income and of pensioners dependent upon state benefits for over 75 per cent of their income are excluded. There is a special pensioner index for the latter group, which is used for upgrading the state pension each year (*Social Trends* 1986, p.209).

The weight (out of 1,000) within the index for different goods varies, so that for instance, mortgage interest payments had a weight of 25 in 1975, but 54 in 1986 (*Building Society News*, May 1986, p.4). Because only about 5 per cent of low-income households have mortgages, the RPI is an inadequate reflection of the way in which the living expenses of poor households vary. However, it is still the basis on which benefits are increased.

SCOTLAND — The same social security system applies in Scotland, though the housing legislation is different. The variations are often fairly slight, for instance in the wording of the fitness standard. In other cases, Scotland's legislation is brought in before that in England and Wales (poll tax) or after (1990 Building Regulations). The gas and electricity industries are both covered by the same Regulators as for England and Wales.

SEVERE EXPOSURE — The effect of the wind is greater on exposed buildings and this may be roughly differentiated, according to the Institute of Heating and Ventilating Engineers (IHVE 1976, pp.A4-9), as follows:

Sheltered:	up to third floor of buildings in city centres;
Normal:	most suburban and country premises; fourth to eighth floors of buildings in city centres;
Severe:	buildings on the coast or exposed on hill sites; floors above the fifth of buildings in suburban or country districts; floors above the ninth of buildings in city centres.

On severely exposed sites, a 50 per cent increase should be allowed in ventilation losses.

SOCIAL CLASS — See socio-economic groups below.

SOCIO-ECONOMIC GROUPS AND SOCIAL CLASS — The former is a marketing description (groups A-E) loosely based on the latter (classes I-V), which is defined by the Registrar General (see Classification of Occupations 1980). The main difference is in the allocation of pensioners: in the former, low-income pensioners are included in group E, whereas with social classes they are coded by previous occupation or husband's, (Table A.2).

SUPPLEMENTARY BENEFIT (SB) — The predecessor to income support (IS) — see above.

SYSTEM BUILDING — See Building Methods and Materials above.

U VALUES — The U value measures the rate at which heat is lost through $1m^2$ of the element when the temperatures each side differ by 1° Celsius, expressed in Watts per square metre per degree Celsius (W/m² °C). These U values are derived from laboratory tests on dry samples in still air. If a material absorbs moisture, whether from rain or from condensation, there is a reduction in its insulation value because water is a better conductor of heat than air. 'The thermal conductivity of masonry may be twice as great when a wall is wet than when it is dry' (Loudon 1980, p.153). The effect on heat loss depends on how long it takes the building to dry out and what proportion of the fabric is affected, but the effect of dampness from rain is probably no more than an additional 10 per cent heat loss from the whole building (IHVE 1975, A3-4). The effect of

Table A.2 Descriptions of social class and socio-economic groups

Social class	Description	Socio-economic group
I	Professional occupations	A
II	Intermediate occupations (including most managerial and senior administrative)	B
IIIN	Skilled occupations (non-manual)	C1
IIIM	Skilled occupations (manual)	C2
IV	Partly skilled occupations	D
V	Unskilled occupations	E*

Note:
* Includes long-term unemployed (over 6 months), and all households solely dependent on state benefits (pensioners in receipt of an occupational pension are allocated to their former class of employment). Based on *Social Trends* 1988, p.199.

condensation does not appear to have been quantified, but many buildings suffer from severe condensation and thus greater heat loss. In Britain it rains from 5 per cent of the time around London, to 15 per cent of the year in north-west Scotland (BRE 1971, p.1), so that increased heat loss occurs in the wettest regions and in homes with condensation. There is a similar cooling effect from wind. These aspects of building physics are 'notoriously unreliable' (Campbell 1985, p.45), and fluctuate rapidly, so they are non-measurable. Thus, the U value is normative, rather than a measure of actual conditions.

UNFIT FOR HUMAN HABITATION — a legal standard giving the basic criteria to be met if a house is to be suitable for people to live in. The definition was extended (slightly) in July 1990, but has left the local Environmental Health Officer with considerable discretion. For the first time in England and Wales, a fixed heating system should be provided in the main living room, but not elsewhere. The cost of keeping warm is not part of the criteria, as there is no guidance on an appropriate fuel, nor on thermal insulation levels.

USEFUL ENERGY — The amount that is available after the deduction of losses which occur when final users convert delivered energy into space or process heat, motive power or light.

References

Abel-Smith, B. and Townsend, P. (1965), *The poor and the poorest*, Bell, London

ABI (1986), *Insurance statistics 1981-85*, Association of British Insurers, London

ACE (1981), *Domestic energy conservation and the UK economy*, prepared by Economists Advisory Group for the Association for the Conservation of Energy, London

ACE (1987), A *lesson from Denmark*, Association for the Conservation of Energy, London

ACEC (1983), *Fifth report to the Secretary of State for Energy*, Advisory Council on Energy Conservation, Energy Paper 52, Department of Energy, HMSO

Aird, A.(1977), 'Goods and services' in F. Williams (ed.), *Why the poor pay more*, Macmillan, London

Alderson, M. R. (1985), 'Season and mortality', *Health Trends*, vol. 17, pp.87-96

Alexander, G., Warm, P. and Reddish, A. (1984), *Warm and wise — energy matters*, the Open University, Milton Keynes

Allen, B. (1984), 'Troubleshooter', *Building*, 7 December, pp.30-1

AMA (1983), *Defects in housing Part 1: 'non-traditional' dwellings of the 1940s and 1950s*, Association of Metropolitan Authorities, London

AMA (1984), *Defects in housing Part 2: industrialised and system built dwellings of the 1960s and 1970s*, Association of Metropolitan Authorities, London

Anderson, B. R., Clark, A. J., Baldwin, R. and Milbank, N. O. (1985a), *BREDEM — BRE domestic energy model: background, philosophy and description*, Building Research Establishment, Garston, Watford

Anderson, B. R., Clark, A. J., Baldwin, R. and Milbank, N. O. (1985b), *BREDEM: the BRE domestic energy model*, Information Paper 16/85, Building Research Establishment, Garston, Watford

Andrews, K. and Jacobs J. (1990), *Punishing the poor — poverty under Thatcher*, Macmillan, London

Architects' Journal, (1985), *Introduction and complete index to the 1985 building regulations*, Architectural Press, London

Ashley, P. (1983), *The money problems of the poor: a literature review*, Heinemann Educational Books, London

Atkins, W. S. and Partners (1984), *CHP/DH feasibility programme: Stage I*, Department of Energy, London

Audit Commission (1986a), *Managing the crisis in council housing*, HMSO

Audit Commission (1986b), *Improving council house maintenance*, HMSO

Bagshaw, M. (1981), 'Domestic energy conservation and the consumer', unpublished M. Phil., University of Bradford

Baillie, A., Cody, C., Griffiths, I. and Huber, J. (1986), 'Domestic energy consumption: houses, people, and comfort', in Monnier *et al.* (1986), *Consumer behaviour and energy policy*, New York and London

Baldwin, R., Henderson, G., Milbank, N. O. and Shorrock, L. D. (1986), 'Energy efficiency in the UK housing stock', paper presented to the 8th Annual North

American Congress of the International Association of Energy Economists, Boston, Mass., November

Barnett, R. R. (1986), 'An economic appraisal of local energy conservation schemes', *Energy Policy*, October, pp.425-36

Bending, R. and Eden, R. (1984), *UK energy: structure, prospects and policies*, Cambridge University Press, Cambridge

Bentham, G. (1990), 'Poverty, cold weather and the rise in infant mortality in England in 1986', paper presented at the Annual Conference of the IBG, Glasgow, School of Environmental Sciences, UEA

Berthoud, R. (1981), *Fuel debts and hardship*, Policy Studies Institute, London

Berthoud, R. (1984), *The reform of supplementary benefit: working papers*, Policy Studies Institute, London

Berthoud, R. (1989), *Credit, debt and poverty*, Social Security Advisory Committee, Research Paper 1, HMSO

Berthoud, R. and Brown, J. C. with Cooper, S. (1980), *Poverty and the development of anti-poverty policy: the United Kingdom*, a report to the Commission of the European Communities, Policy Studies Institute, London

Berthoud, R. and Kempson, E. (1990), *Credit and debt in Britain*, Policy Studies Institute, London

Black, F. W. (c.1962), *Heating for old people: some conclusions based on their experience of electric floor-warming*, Building Research Establishment, Current Paper, Design Series 19, Watford

Boardman, B. (1984), *The cost of warmth: an energy efficient approach to overcoming fuel poverty*, discussion paper, National Right to Fuel Campaign, Birmingham

Boardman, B. (1985), 'Activity levels within the home', paper presented to the Joint Meeting CIB W17/77, *Controlling Internal Environment*, Budapest, 18-20 September

Boardman, B. (1986a), 'Fuel poverty: the need for a low-income energy efficiency programme', *Energy UK 1986*, eds A Harrison and J Gretton, Policy Journals, Newbury, Berks

Boardman B. (1986b), 'Seasonal mortality and cold homes', paper given at *Unhealthy Housing Conference*, Institution of Environmental Health Officers and Legal Research Institute, University of Warwick, 14-16 December

Boardman B. (1990), *Fuel poverty and the greenhouse effect*, National Right to Fuel Campaign, Heatwise Glasgow, Neighbourhood Energy Action, Friends of the Earth

Bordass, W. (1984), *Heating your church*, C10 Publishing, London

Bowdidge, J. R. (undated — 1989/90), *Pollution reduction through energy conservation*, Eurisol, Redbourn, Herts

Bowers, J. (1990), *Economics of the environment*, British Association of Nature Conservationists, Telford, Shropshire

Boyd, D., Cooper, P. and Oreszczyn, T. (1988), 'Condensation risk prevention: additions of a condensation model to BREDEM', *Building Services Engineering Research and Technology*, vol 9, No 3, pp.117-25

Bradshaw, J. (1980), 'Cold conditions — the social circumstances', paper given to *Cold Conditions Conference*, National Fuel Poverty Forum, London, December 1979

Bradshaw, J. and Hardman, G. (1988), *Expenditure on fuels 1985*, Gas Consumers' Council, London

Bradshaw, J., Hardman, G. and Hutton, S. (1986), *Expenditure on fuels 1983*, published jointly by National Gas Consumers' Council and Electricity Consumers' Council, London

Bradshaw, J. and Harris, T. (eds.) (1983), *Energy and social policy*, Routledge & Kegan Paul, London

Bradshaw, J. and Hutton, S. (1983), 'Tariff tilting' in J. Bradshaw and T. Harris (eds.), *Energy and social policy*, Routledge & Kegan Paul,London

Bravery, A. F., Grant, C. and Sanders, C. H. (1987), 'Controlling mould growth in housing', paper presented to *Unhealthy Housing: Prevention and Remedies Conference*, Institution of Environmental Health Officers and Legal Research Institute, University of Warwick, 13-15 December

BRE (1971), *An index of exposure to driving rain*, Digest 127, Building Research Establishment, Watford

BRE (1979), *Thermal, visual and acoustic requirements in buildings*, Digest 226, Building Research Establishment, Watford

Britain 1975: an official handbook, HMSO

British Medical Association (1987), *Deprivation and ill-health*, paper for discussion, Board of Science and Education, London

British Medical Journal (1964), 'Cold the killer' and 'Accidental hypothermia in the elderly', 14 November, pp.1212-13 and pp.1255-8

Brown, I. (1990), *Least-cost planning in the gas industry*, Ofgas, London

Brundrett, G. W. (1987), *Living in highly insulated buildings*, Electricity Council Research Centre, Capenhurst, Chester

BSI (1986), 'British Standard Code of Practice for Energy Efficiency in Housing — Draft', British Standards Institution, London, 26 September

BSRIA (1983), *Systems profile*, Building Services Research and Information Association, Bracknell, June

Bull, D. (1971), 'The rediscovery of family poverty' in D. Bull (ed.), *Family poverty: programme for the seventies*, Duckworth, London

Burghes, L. (1980), *Living from hand to mouth*, Family Services Units and Child Poverty Action Group, London

Bunn, D. and Vlahos, K. (1989), 'Evaluation of the long-term effects on UK electricity prices following privatisation', *Fiscal Studies*, Vol 10, No 4, November, pp.104-16

Burr, M. L. (1986), 'Damp housing and respiratory disease', paper presented to *Unhealthy Housing Conference*, Institution of Environmental Health Officers and Legal Research Institute, Warwick, 14-16 December

Burton, S. and Hughes, D. (1985), *Lea View — low energy*, LEEN Report 3, ECSC, London

Byrd, H. (1986), 'Hard to heat houses', *Housing*, September pp.23-4

Byrne, D. S., Harrison, S. P., Keithley, J. and McCarthy, P. (1986), *Housing and health: the relationship between the housing conditions and the health of council tenants*, Gower, Aldershot

Campbell, P. M. (1985), *The better insulated house programme*: Reference Report for the Department of the Environment, prepared by Databuild Ltd., DOE, London

CEGB (1988), *Medium and long-term load estimates: methodology and forecasts*, 1987/88 planning cycle, London

Census 1981, National Report GB Pt 1, HMSO

Chalker, L. (1980), 'Introductory address', *Cold Conditions Conference*, National Fuel Poverty Forum, London, December 1979

Chapman, P. (1975), *Fuel's paradise*, Penguin Special, Harmondsworth, Middx.

Chateau, B. and Lapillone, B. (1982), *Energy demand: facts and trends*, Springer-Verlag, Vienna

Chesshire, J. (1986), 'An energy-efficient future — a strategy for the UK', *Energy Policy*, October, pp.395-412

Chesshire, J. H., Friend, J. K., Pollard, J. de B. Stringer, J. and Surrey, A. J. (1977), 'Energy policy in Britain: a case study of adaptation and change in a policy system', in L. N. Lindberg (ed.), *The energy syndrome*, Lexington Books, Lexington, Mass., USA

Chesshire, J. H. and Surrey, A. J. (1978), *Estimating UK energy demand for the year 2000: a sectoral approach*, Science Policy Research Unit, Occasional Paper 5, University of Sussex, Brighton

CIBS Guide A1 (1978), Chartered Institution of Building Services, London

Circuit News (monthly), Electricity Council, London

Claxton, J. D., Gorn, G. J., Weinberg, C. B. (1986), 'Energy policy to serve low income households', in Monnier *et al.* (1986), *Consumer behaviour and energy policy*, Praeger, New York and London

Cmd 6762 (1946), *Domestic fuel policy*, Fuel and Power Advisory Council, Ministry of Fuel and Power, HMSO, London (The Simon Report)

Cmd 8647 (1952), *Report of the Committee on the use of national fuel and power resources*, HMSO (The Ridley Committee)

Cmnd 6615 (1976), *Supplementary Benefits Commission annual report 1975*, HMSO

Cmnd 9513 (1985), *Home improvement — a new approach*, HMSO (Green Paper)

Cmnd 9518 (1985), *Reform of social security — programme for change*, vol. 2, HMSO

Cmnd 9519 (1985), *Reform of social security*, vol. 3, HMSO

Cmnd 9520 (1985), *Housing benefit review*, Report of the Review Team, HMSO

Cmnd 801 (1989), *Developments in the European Community January-June 1989*, House of Commons, HMSO

Coates, K. and Silburn, R. (1970), *Poverty: the forgotten englishmen*, Penguin, Harmondsworth, Middx.

Cohen, R. and Lakhani, B. (1986), *National welfare benefits handbook*, Child Poverty Action Group, London

Cold homes: the crisis (1982), The Report of the National Conference, National Right to Fuel Campaign, Bradford, 20 May

Collins, K. J. (1986), 'The health of the elderly in low indoor temperatures', paper presented to *Unhealthy Housing Conference*, Institution of Environmental Health Officers and Legal Research Institute, Warwick, 14-16 December

Collins, K. J. and Exton-Smith, A. N. (1983), 'Thermal homeostatis in old age', *Journal of the American Geriatrics Society*, vol. 31, no. 9, September, pp.519-24

Collins, K. J. (1989), *Hypothermia and seasonal mortality in the elderly. Care of the elderly*, vol. 1, no. 6, November, pp.257-9

Conaty, P. (1987), *Birmingham Settlement money adviser*

Cook, P. L. and Surrey, A. J. (1977), *Strategies for uncertainty*, Martin Robertson, London

Cooper, I. (1982), 'Energy conservation in buildings — Part 2: a commentary on British Government thinking', *Applied Energy*, 10, 1, pp.1-45

Cooper, S. (1981), *Fuel poverty in the United Kingdom*, prepared for the Commission of the European Communities, Policy Studies Institute, London

Cornish, J. P. (1977), 'The effect of thermal insulation on energy consumption in houses' in R. Courtney (ed.), *Energy conservation in the built environment*, Construction Press, Watford

CPAG (1974), *Cold comfort*, Child Poverty Action Group, London

CPAG (1980), *Living from hand to mouth*, joint report from Child Poverty Action Group and Family Services Units, London

Crawshaw, A. J. E., Williams, D. I. and Crawshaw, C. M. (1985), 'Consumer knowledge and electricity consumption', *Journal of Consumer Affairs and Home Economics*, 9, pp.283-9

Crawshaw, C. M. and Dale, H. C. A. (1981), 'Users' understanding of a heating system' paper delivered to CIB-5-17 Conference, *Heating and Climatisation*, Delft

Croome, D. J. and Roberts, B. M. (1981), *Airconditioning and ventilation of buildings*, 2nd ed., Pergamon Press, Oxford

Cullingworth, J. B. (1966), *Housing and local government in England and Wales*, George Allen & Unwin, London

Curwen, M. (1981), *Trends in respiratory mortality 1951-75, England and Wales*, OPCS DH1 no.7, HMSO

Curwen, M. and Devis, T. (1988), 'Winter mortality, temperature and influenza: has the relationship changed in recent years?', *Population Trends 54*, OPCS pp.17-20

Danish Building Agency (1986), *Denmark uses energy better*, Copenhagen

Darmstadter, J. and Edmonds, J. (1989), 'Human development and carbon dioxide emissions: the current picture and the long-term prospects' in N. J. Rosenberg, W. E. Easterling III, P. R. Crosson and J. Darmstadter (eds.), *Greenhouse warming: abatement and adaption*, Workshop proceedings, 14-15 June 1988, Resources for the future, Washington DC, USA

Davidson, P. (1990), 'How to meet the new thermal regs 1', *Building Today*, 18 January, pp.16-17

Davies, D. T. I. G. (1986), 'An overview of the code, its aims and objectives', in *Energy — waste not, want not*, summary of speakers' presentations, conference on *Code of Practice for Energy Efficiency in Housing*, organized by the British Standards Institution, Wembley, 5 November

Deacon, A. and Bradshaw, J. (1983), *Reserved for the poor: the means test in British social policy*, Basil Blackwell and Martin Robertson, London

DEn (1976a), *Energy tariffs and the poor*, Department of Energy, London

DEn (1976b), *National Energy Conference*, Energy Paper 13, vol. I, Report of Proceedings, Department of Energy, HMSO

DEn (1976c), *Review of payment and collection methods for gas and electricity bills*, report of an Informal Inquiry (The Oakes Report), Department of Energy

DEn (1978a), *Family expenditure survey: expenditure on fuels 1976*, Department of Energy, London

DEn (1978b), *Energy forecasting methodology*, Energy Paper 29, Department of Energy, HMSO

DEn (1982a), *Speech on energy policy by the Secretary of State for Energy, at Cambridge, 28 June*, Energy Paper 51, Department of Energy, HMSO

DEn (1982b), *Advisory Council on Energy Conservation*, report to the Secretary of State for Energy, Energy Paper 49, HMSO

DEn (1983a), *Government observations on the Fifth Report from the Select Committee on Energy — Session 1981-82*, Department of Energy, London

DEn (1983b), *Investment in energy use as an alternative to investment in energy supply*, DEN/S/3 (NE), Department of Energy, London, January

DEN 1 (1977), *Domestic energy notes 1: selection criteria for electric space and water heating systems in new dwellings*, Joint Working Party on Heating and Energy Conservation in Public Sector Housing, now available from the Department of the Environment, London

DEN 3 (1978), *Domestic energy notes 3: remedial work for existing electrically heated dwellings*, Joint Working Party on Heating and Energy Conservation in Public Sector Housing, now available from Department of the Environment, London

DHSS (1972), *Keeping warm in winter*, Department of Health and Social Security leaflet, September

Dickens, P., Duncan, S., Goodwin, M. and Gray, F. (1985), *Housing, states and localities*, Methuen, London

Dilnot, A. and Helm, D. (1987), 'Energy policy, merit goods and social security', *Fiscal Studies*, vol. 8, no. 3, Institute for Fiscal Studies, London, August

Dinan, T. M. and Trumble, D. (1989), 'Temperature takeback in Hood River conservation project', *Energy and Buildings*, 13, pp.39-50

DOE (1973), *Thermal insulation in housing*, Housing Development Notes, Department of the Environment, HMSO, December

DOE (1978), *An exploratory project on heating for the elderly*, Housing Development Directorate, Department of the Environment, London

DOE (1983), *English house condition survey*, Part 2, Department of the Environment, HMSO,

DOE (1985), *An inquiry into the condition of local authority housing stock in England*, Department of the Environment, HMSO

DOE (1986a), *Housing and construction statistics 1975-85*, HMSO

DOE (1986b), *Review of the Building Regulations*, Department of the Environment, 15 December

DOE (1986c), *Housing and construction statistics: Great Britain*, September Quarter, Part 1, Department of the Environment, HMSO

DOE (1986d), *Energy efficient renovation of houses, a design guide*, Department of the Environment, HMSO

DOE (1988), *1986 English house condition survey*, HMSO

DOE (1989a), *Housing and construction statistics, Great Britain*, September Quarter, part 2, HMSO

DOE (1989b), *Housing and construction statistics, Great Britain*, December quarter FB, HMSO

DOE (1989c), *Housing and construction statistics 1977-87*, HMSO

DOE (1990a), *The appraisal of local authority housing — the condition of the local authority housing stock in England: 1988*, Department of the Environment, London

DOE (1990b), *Estate action — annual report 1988-1989*, Department of the Environment, HMSO

Donnison, D. (1982), *The politics of poverty*, Martin Robertson, Oxford

DUKES (annually), *Digest of United Kingdom Energy Statistics*, Department of Energy, HMSO. The year quoted in the reference is the year of the survey, rather than of publication

Dunleavy, P. (1981), *The politics of mass housing in Britain, 1945-75*, Clarendon Press, Oxford

Durward, L. (1981), *Elderly electricity consumers*, Research Report 4, Electricity Consumers' Council, London

Economic Trends (monthly), Central Statistical Office, HMSO

EEDS (1983), *External wall insulation applied to 'Woolaway' system houses*, Energy Efficiency Demonstration Scheme, no. F/27/83/88, Tewkesbury, Building Research Energy Conservation Support Unit, Watford

EEDS (1984), *Low energy houses in the city of Manchester*, Energy Efficiency Demonstration Scheme, Building Research Energy Conservation Support Unit, Watford

EEDS 56 (1988), *Energy advice to tenants*, Energy Efficiency Demonstration Scheme, R & D 56, BRECSU, Watford

EEO (1983), *A guide to home heating costs*, Energy Efficiency Office, HMSO. Guides were produced for the four main regions: South, Midlands and North, Wales, Scotland

EEO (1984), *Degree days*, Fuel Efficiency Booklet 7, Energy Efficiency Office, HMSO

EEO (1986), *Cutting home energy costs: a step-by-step monergy guide*, four editions, one each for gas, electricity, solid fuel and oil users, Energy Efficiency Office, London

EIK Project (1980), *Report on Phase I*, City of Birmingham Housing Department and Department of the Environment, London

EIK Project (1982), *Report on Phase II*, vol. 1, City of Birmingham Housing Department and Department of the Environment, London

Elder, A. J. (1977), *Guide to the Building Regulations 1976*, Architectural Press, London

Electricity Consumers' Council (1985), *Debt collection, disconnections and electricity consumers: report on the operation of the code of practice*, discussion paper 14, London

Electricity Council (1963), *Report and accounts for the year ended 31 March 1963*, HMSO

Electricity Council (1989a), *Handbook of electricity supply statistics*, London

Electricity Council (1989b), *Electricity supply statistics*

Electricity Council (1989c), *Tariffs review*, London

Electricity Council (undated), *Direct electric heating design manual*, London

Energy Action Bulletin, produced four times a year by Neighbourhood Energy Action, Newcastle

Energy in Europe (1985), 'Energy saving in Denmark', December, pp.25-8

Etheridge, D. W. and Nevrala, D. J. (1978), 'Air infiltration in the UK and its impact on the thermal environment' in *Indoor climate: effects on human comfort, performance and health in residential, commercial and light industry buildings*, WHO Conference, Copenhagen

Eurisol (1984), *Building Regulations and domestic insulation*, Insulation Fact Sheet no. 18, Bromley, Kent

Evans, R. D. and Herring, H. P. J. (1989), *Energy use and energy efficiency in the UK domestic sector up to the year 2010*, EEO II , HMSO

Fairmaner, W. D. (1988), *Periods of exceptionally severe weather: towards a definition?*, (Working paper 39) Dept. of Geog, University of Birmingham

Fanger, P. O. (1973), 'Conditions for thermal comfort — a review' in *Thermal comfort and moderate heat stress*, proceedings of the CIB Commission W45 (Human Requirements) Symposium, held at Building Research Station 13-15 September 1972, BRE, HMSO

FES (annually), *Family Expenditure Survey*, Department of Employment, HMSO. The year quoted in the reference is the year of the survey, rather than of publication.

Field, J. and Hedges, B. (1977), *National fuel and heating survey*, Social and Community Planning Research, prepared for the National Consumer Council, London

Finer, E. G. (1982), *How the Government handles energy conservation*, Rayner Scrutiny, Department of Energy, London

Fishman, D. S. and Pimbert, S. L. (1982), 'The thermal environment in offices', *Energy and Buildings*, 5, pp.109-16

Fisk, D. J. (1977), 'Microeconomics and the demand for space heating', *Energy*, vol. 2, pp.391-495

Forrest, R. and Murie, A. (1987), 'The pauperisation of council housing', *Roof*, January/February, pp.20-3

Franey, R. and Reason, L. (1985), *Electric heating in public sector housing*, Research Report 14, Electricity Consumers' Council, London

Fry, V. and Pashardes, P. (1986), *The RPI and the cost of living*, Institute for Fiscal Studies, Report Series no. 22, London

Gershuny, J., Miles, I., Jones, S., Mullings, C., Thomas, G. and Wyatt, S.(1986),

'Time budgets: preliminary analyses of a national survey', *Quarterly Journal of Social Affairs*, 2 (1), pp.13-39

GHS (annually), *General Household Survey*, Office of Population Censuses and Surveys, HMSO. The year quoted in the reference is the year of the survey, rather than the year of publication.

Gibson, M. and Price, C. (1986), 'Standing charge rebates: costs and benefits', *Energy Policy*, June, pp.262-71

GLC (1986), *Housing standards: a survey of new build local authority housing in London 1981-1984*, Greater London Council, London

GLCABS (1984), *Cold comfort for the poor: a survey of fuel debt problems in London*, Greater London Citizens Advice Bureaux Service, London

Golding, P. (ed.) (1986), *Excluding the poor*, CPAG, London

Goode, J., Roy, D. and Sedgewick, A. (1980), *Energy Policy: a reappraisal* Fabian Research Series 343, London

Government Statistical Service (1990), *Households below average income — a statistical analysis 1981-87*, Department of Social Security, London

Gray, M., Johnson, M., Seagrave, J., Dunne, M. (1977), *A policy for warmth*, Fabian Tract 447, Fabian Society, London

Gray, P. G. and Russell, R. (1962), *The housing situation in 1960*, Central Office of Information, London

Green, J., Innes, W., Maby, C. and Osbaldeston, J. (1987), *Heating advice handbook*, Energy Inform, Newcastle

Haskey, J. (1986), 'One-parent families in Great Britain' in *Population trends 45*, Office of Population Censuses and Surveys, London

HC 353 (1976), *Gas and electricity prices, fourth report*, Session 1975-6, Select Committee on Nationalised Industries, House of Commons, HMSO

HC 352-ii (1981), *Energy conservation: buildings*, Minutes of Evidence 15 June 1981, Fifth Report, Session 1981-2, Select Committee on Energy, HMSO

HC 401-I (1982), *Energy conservation in buildings*, vol. 1 Report and Minutes of Proceedings, Fifth Report, Session 1981-2, Select Committee on Energy, HMSO

HC 401-II (1982), *Energy conservation in buildings*, vol. 2, Fifth Report, Session 1981-2, Select Committee on Energy, HMSO

HC 402 (1982), *Energy conservation in buildings*, Fifth Report, Session 1981-2, Select Committee on Energy, House of Commons, HMSO

HC 276 (1984), *Electricity and gas prices*, First Report, Session 1983-4, Energy Committee, House of Commons, HMSO

HC 87 (1985), *The energy efficiency office*, Eighth Report, Session 1984-5, Energy Committee, House of Commons, HMSO

HC 262 (1986), *The Government's response to the Committee's Eighth Report (Session 1984-85) on the Energy Efficiency Office*, Third Report, Session 1985-6, Energy Committee, House of Commons, HMSO

HC 54 (1988) *Perinatel, neonatal and infant mortality*, First Report, Session 1988-9, Social Services Committee, House of Commons, HMSO

HC 292 (1989) *Energy policy implications of the greenhouse effect*, Sixth Report, Session 1988-89, Energy Committee, Vol I, House of Commons HMSO

HC 378-II (1990), *The income support system and the distribution of income in 1987*, Social Services Committee, Session 1989-90, House of Commons, HMSO

HC 395 (1990), *National energy efficiency*, Fourteenth Report, Committee of Public Accounts, Session 1989-90, HMSO

HC 405-i (1990), *Energy efficiency*, Select Committee on Energy, Session 1989-90, Minutes of evidence, House of Commons, HMSO

Health Education Authority (1987), *The health divide: inequalities in health in the 1980's*, a review prepared by M. Whitehead, London

Henderson, G. (1986), 'Energy efficiency in UK housing', *Building Services*, January

Henderson, G. and Shorrock, L. D. (1989), *Domestic energy fact file*, BRE, HMSO

Henderson, G. and Shorrock, L. D. (1990), *Greenhouse-gas emissions and buildings in the United Kingdom*, IP2/90, BRE Information Paper, Watford

Hesketh, J. L. (1973), *Social problems associated with local authority central heating*, Family Welfare Association of Manchester and Salford

Hesketh, J. L. (1978), *Inside the system: how an Electricity Board deals with fuel debtors*, Family Welfare Association of Manchester

Heslop, D. T. and Sussex, A. D. (1984), 'The new housing challenge — the reality', *New Housing*, no. 8, Studies in Energy Efficiency in Buildings, British Gas, London

Hillman, M. and Bollard, A. (1985), *Less fuel, more jobs*, Policy Studies Institute, London

Housing Corporation (1985), *Design and contract criteria*, Issue 3/3, London

Humphreys, M. A. (1976), *Desirable temperatures in dwellings*, CP 75/76, Building Research Establishment, Watford

Humphreys, M. A. (1978), 'The influence of season and ambient temperature on human clothing behaviour' in *Indoor Climate*, proceedings of a Conference, Copenhagen, World Health Organization

Hunt, D. R. G. and Gidman, M. I. (1982), 'A national field survey of house temperatures', *Building and Environment*, vol. 17, no. 2, pp.107-24

Hunt, D. R. G. and Steele, M. R. (1980), 'Domestic temperature trends', *The Heating and Ventilating Engineer*, April, pp.5-15

Hunt, S. M., Martin, C. J. and Platt, S. D. (1986), 'Housing and health in a deprived area of Edinburgh', paper presented to *Unhealthy Housing — A Diagnosis Conference*, Institution of Environmental Health Officers and Legal Research Unit, University of Warwick, 14-16 December

Hutton, S., Bradshaw, J. and Hardman, G. (1987), 'Domestic fuel expenditure and payment of fuel allowances', *Journal of Consumer Studies and Home Economics*, vol. 11, pp.1-20

Hutton, S., Gaskell, G., Pike, R., Bradshaw, J. and Corden, A. (1985), *Energy efficiency in low income households: an evaluation of local insulation projects*, Energy Efficiency Office 4, HMSO

Hutton, S. and Hardman, G. (1990), *Expenditure on fuels 1986*, Gas Consumers Council (in press)

IEA (1987), *Energy conservation in IEA Countries*, International Energy Agency, OECD, Paris

IEA (1989), *Electricity end-use efficiency*, OECD, Paris

IEHO (1986), *Unhealthy housing — a diagnosis*, proceedings of a conference held by the Institution of Environmental Health Officers and the Legal Research Institute, University of Warwick, 14-16 December

IHVE (1975), *Guide A3: thermal and other properties of building*, Institute of Heating and Ventilating Engineers, London

IHVE (1976), *Guide A4: air infiltration*, Institute of Heating and Ventilating Engineers, London

Institute of Housing, and Royal Institute of British Architects (1983), *Homes for the Future*, the Institute and RIBA

Isaacs, N. and Donn, M. (1990), 'Seasonality in New Zealand mortality', *NZ Med J.* (in press)

Isherwood, B. C. and Hancock, R. M. (1979), *Household expenditure on fuel: distributional aspects*, Economic Adviser's Office, DHSS, London

Jackson, T. and Roberts, S. (1989), *Getting out of the greenhouse*, Friends of the Earth, London

Jones, C. (1989), 'evidence to House of Lords Select Committee on the European Communities', report HL 37, *Efficiency of Electricity Use*, 8th report, Session 1988-89, HMSO

Keatinge, W. R. (1986), 'Seasonal mortality among elderly people with unrestricted home heating', *British Medical Journal*, 293, 20 September, pp.732-3

Keatinge, W. R., Coleshaw, S. R. K. and Holmes, J. (1989), 'Changes in seasonal mortalities with improvement in home heating in England and Wales from 1964 to 1984', *Int. J. Biometeorol*, 33, pp.71-6

Keatinge, W. R., Coleshaw, S. R. K., Cotter, F., Mattock, M., Murphy, M. and Chelliah, R. (1984), 'Increases in platelet and red cell counts, blood viscosity, and arterial pressure during mild surface cooling. Factors in mortality from coronary and cerebral thrombosis in winter', *British Medical Journal*, 289, pp.1405-8

Kendrew, W. G. (1961), *The climates of the continents*, 5th edn. Clarendon Press, Oxford

Landy, M. P. (1985), *A computerised energy audit for low income housing*, M.Sc. dissertation, University of Sussex

Leach, G., Lewis, C., Romig, F., van Buren, A. and Foley, G. (1979), *A low energy strategy for the UK*, International Institute for the Environment and Development, London

Leach, G. and Pellew, S. (1982), *Energy conservation in housing*, International Institute for the Environment and Development, London

Legal Action Bulletin, newsletter of the Legal Action Group, London

Levitt, D. and Burrough, A. (1979), 'Rehab at all costs', *Architects'Journal*, Architectural Press, London, 4 July, pp.17-29

Lewis, P. (1982), *Fuel poverty can be stopped*, National Right to Fuel Campaign, Bradford

Lipsey, R. G. (1979), *An introduction to positive economics*, 5th edn., Weidenfeld and Nicolson, London

Loader, P. T. and Milroy, E. A. (1961), 'Space and water heating in local authority flats, 1', *Architects' Journal*, 1 June, pp.807-14

Lorant, J. (1981), *Poor and powerless: fuel problems and disconnections*, Poverty Pamphlet 52, Child Poverty Action Group, London

Lorber, S., Pierce, S., Tysh, I. and Winter, J. (1989), *Fuel rights handbook*, SHAG/WRUG, London

Loudon, A. (1980), 'Concrete blockwork', *Architects' Journal*, 10 December

Lowe, R., Chapman, J. and Everett, R. (1985), *The Pennyland project, final report*, Energy Research Group, Open University, Milton Keynes

McGeevor, P. A. (1982), 'The active pursuit of comfort: its consequences for energy use in the home', *Energy and Buildings*, 5, pp.103-7

Macintyre, W. (1985), *Evidence to the Select Committee on Energy inquiry into the Energy Efficiency Office*, HC 87, Eighth Report, Session 1984-85, HMSO, 12 June

McKee, C. M. (1989), 'Deaths in winter: can Britain learn from Europe?', *European Journal of Epidemiology*, vol. 5, no. 2, June, pp.178-82

McKee, C. M. (1990), 'Deaths in winter in Northern Ireland: the role of low temperature', *Ulster Medical Journal*, vol. 59, no. 1, April, pp.17-22

MacKerron, G. (1987), Energy pricing: industrialised country practice versus criteria for developing countries', paper presented to the Third World Energy Policy Study Group Workshop, Kings College, London, 1 October

McNair, H. P. (1986), *British Gas*, Watson House, pers comm

Mack, J. and Lansley, S. (1985), *Poor Britain*, George Allen & Unwin, London

Mant, D. C. and Muir Gray, J. A. (1986), *Building regulation and health*, Building Research Establishment Report, HMSO

Markus, T A (1982), 'Development of a cold climate severity index', *Energy and Buildings*, 4, pp.277-83

Markus, T. A. (1987a), Cold, condensation, climate and poverty in Glasgow', paper presented at the *Unhealthy Housing: Prevention and Remedies Conference*, Institution of Environmental Health Officers and Legal Research Institute, University of Warwick, 13-15 December

Markus, T. A. (1987b), 'Heat with rent: a study of performance', unpublished report, with restricted circulation

Markus, T. A. and Morris, E. N. (1980), *Buildings, climate and energy*, Pitman, London

Markus, T. A. and Nelson, I. (1985), *An investigation of condensation dampness*, Research Project Report, Department of Architecture and Building Science, University of Strathclyde, Glasgow

MAS (1979), *Electricity users survey — summary report*, Electricity Consumers' Council, London

Merrett, S. (1979), *State housing in Britain*, Routledge & Kegan Paul, London

Meyel, A. (1987), *Low income households and energy conservation*, Built Environment Research Group, Polytechnic of Central London, London

Milbank, N. O. (1986a), 'The potential for passive solar energy in UK housing', paper presented at the CIB W67 meeting, Lisbon, June

Milbank, N. O. (1986b), 'Is there a minimum heating requirement for households?', *Building Services Engineering Research & Technology*, vol. 7, no. 1, pp.44-6

Milbank, N. O., Cornish, J. P., Sanders, C. H. and Garratt, J. (1985), 'User behaviour and the effectiveness of remedial treatments for surface condensation and mould', paper presented to the Joint Meeting CIB W17/77, Controlling Internal Environment, Budapest, 18-20 September

Mintel (monthly), Market Intelligence Reports, London

MOHLG (1961), *Homes for today and tomorrow*, Ministry of Housing and Local Government, HMSO (Parker Morris Report)

MOHLG (1968), *Some aspects of designing for old people*, Design Bulletin 1, HMSO

Monergy News 2 (1987), newsletter produced by the Energy Efficiency Office, London

Mongar, A. (1986), British Gas, Rivermill House, pers comm

Monnier, E., Gaskell, G., Ester, P., Joerges, B., LaPillone, B., Midden, C. and Puiseux, L. (1986), *Consumer behaviour and energy policy: an international perspective*, Praeger, New York and London

Moore, R. (1987), 'The development and role of standards for the older housing stock', paper presented to the Conference on *Unhealthy Housing: Prevention and Remedies*, University of Warwick, Legal Research Unit and Institution of Environmental Health Officers, 13-15 December

Murie, A. (1983), *Housing inequality and deprivation*, Heinemann, London

Murphy, M. F. G. and Campbell, M. J. (1987), 'Sudden infant death syndrome and environmental temperature: an analysis using vital statistics', *J. of Epidemiology and Community Health*, 41, pp.63-71

Muthesius, H. (1979), *The English house*, Granada, London (written in 1904 in German, translated in 1979)

Muthesius, S. (1982), *The English terraced house*, Yale University Press, London

NACAB (1983), *Poverty and paying for fuel: the CAB experience*, National Association of Citizens Advice Bureaux, London

NACAB (1986), *Report of the Money Advice Project Group*, National Association of Citizens Advice Bureaux, London, January

NAO (1989), *National energy efficiency*, Report by the Comptroller and Auditor General, HC 547, HMSO

National Council for One Parent Families (1978), *Fuel debts and one-parent families*, London

National Council for One Parent Families (1985), *Fuel poverty: case-studies from one parent families*, London

Nayha, S. (1984), 'The cold season and deaths in Finland', *Arct. Med. Res.*, 37, pp.20-4

NBA Tectonics (undated), *Energy conservation in housing rehabilitation*, Briefing Note no. 7, National Federation of Housing Associations, London

NCC (1976), *Paying for fuel*, National Consumer Council, Report Number 2, HMSO

NCC (1979), *Soonest mended: a review of the repair, maintenance and improvement of council housing*, National Consumer Council, London

NCC (1982), *Cracking up — building faults in council homes: proposals for a New Deal*, London

NCC (1990), *Credit and debt: the consumer interest*, National Consumer Council, HMSO

NEDO (1974), *Energy conservation in the United Kingdom: achievements, aims and options*, National Economic Development Office, HMSO, London

Nevrala, D. J. (1979), 'The effect of insulation, mode of operation and air leakage on the energy demand of dwellings in the UK' in *Ventilation of domestic Buildings*, British Gas, London, June 1981

NFHA (1985), *New lettings survey*, National Federation of Housing Associations, London

Nicholls, E. and Rees, A. M. (1983), 'Energy utilisation of appliances, energy auditing and energy conservation', paper presented at *The Fifth Home Economics Research Conference*, 15-16 September, organized by School of Home Economics, University College, Cardiff

Nottingham Heating Project (1985a), *Drifting away?, a pilot report on heating standards in the Caunton Avenue Flats*, Nottingham

Nottingham Heating Project (1985b), *The cruel cost of coldness*, Nottingham

NRFC (1977), *Supply and demand: the policies of the National Right to Fuel Campaign*, National Right to Fuel Campaign, Birmingham

NRFC (1982), *Cold homes the crisis*, report of a national conference, National Right to Fuel Campaign, Birmingham

NRFC (1987), *Winds of benefit changes blow colder: heating additions and income support — winners and losers*, National Right to Fuel Campaign, Birmingham

Olivier, D. (1985), 'There's a lot to learn from the Scandinavians' in *What's New in Building*, January

Olivier, D., Miall, H., Nectoux, F. and Opperman, M. (1983), *Energy-efficient futures: opening the solar option*, Earth Resources Research, London

Optima (1989), *Neighbourhood heating advisers*, Optima Energy Services, London

Ormandy, D. (1986), 'Fixing It' *Roof*, May/June, pp.15-18

Ouseley, R. (1988), *An analysis of gas disconnections*, Gas Consumers' Council, London

Owen. G. (1989a) *Room for improvement*, Policy Discussion Paper 1, Neighbourhood Energy Action, Newcastle

Owen, G. (1989b), *Access to improvement*, Policy Discussion Paper 2, Neighbourhood Energy Action, Newcastle

Pahl, R. E. (1985), 'The social and political implications of household work strategies', *Quarterly Journal of Social Affairs*, 1 (1), pp.9-18

Parker, G. (1985), 'Consumers in debt', a background paper for the National Consumer Council Conference *Consumers in Debt*, 14 January 1986, London

Pearson, M. and Smith, S. (1990), *Taxation and environmental policy: some initial evidence*, Institute for Fiscal Studies, Commentary 19, London

Penz, F. (1983), *Passive solar heating in existing dwellings*, Martin Centre for Architectural and Urban Studies, University of Cambridge

Pezzey, J. (1984), *An economic assessment of some energy conservation measures in housing and other buildings*, BRE, Watford

Pimbert, S. and Fishman, D. (1982), 'How warm do people run their homes?', *Building Services & Environmental Engineer*, October, pp.5-7

Plumpton, A. (1984), 'From costs to tariffs', in *Basis of electricity tariffs in England and Wales*, Electricity Council, London

Poverty (quarterly), Child Poverty Action Group Journal, London

Priddle, R. J. (1982), *Proof of evidence for the Sizewell 'B' Public Inquiry*, Department of Energy, London

RIIA and SPRU (1989), *A single European market in energy*, Royal Institute for International Affairs, London

Rollett, C. (1972), 'Housing' in A. H. Halsey (ed.), *Trends in British society since 1900*, Macmillan, London and Science Policy Research Unit, University of Sussex, Brighton

Sakamoto-Momiyama, M. (1977), *Seasonality in human mortality*, University of Tokyo Press, Tokyo

Schipper, L., Meyers, S., Kelly, H. and Associates (1985), *Coming in from the cold: energy-wise housing in Sweden*, Seven Locks Press, Washington, DC, USA

Shaw, J. (1971), *On our conscience: the plight of the elderly*, Penguin, Harmondsworth, Middx.

Sheffield, City of (1986), *Heating in tower blocks: an assessment of Economy 7*, Housing Information, Sheffield, October

Sheldrick, B. (1985), *Hard-to-heat estates: a review of policy and practice*, DHSS 267 BS.10/85, Social Policy Research Unit, University of York, York

Sheldrick, B. (1986) 'Hard-to-heat estates — the need for urgent action', *Energy Action Bulletin*, Neighbourhood Energy Action, Newcastle

Shorrock, L. D. and Henderson, G. (1990), *Energy use in buildings and carbon dioxide emissions*, Building Research Establishment, Garston, Watford.

Simmonds, P. (1987), *UK energy policy: fuel poverty and the need for a low-income fuel substitution programme*, M.Sc. dissertation, Science Policy Research Unit, University of Sussex, Brighton

Sinfield, A. and Fraser, N. (1985), *The real cost of unemployment*, Dept. of Social Administration, University of Edinburgh

Slesser, M. (Ed.) (1982), *Dictionary of energy*, Macmillan Reference Books, London

Smail, R. (1985), 'The price of low pay — deepening poverty', *Low Pay Review*, Winter 1985-86, no. 24, Low Pay Unit

Smith, K. (1986), *'I'm not Complaining' — the housing conditions of elderly private tenants*, Kensington and Chelsea Staying Put for the Elderly in Association with SHAC

Smith, M. E. H., (1977), *Guide to housing*, Housing Centre Trust, London

Social Policy Research Unit (1981), *The reliability of data on household fuel expenditure in the family expenditure survey*, Fuel 7: 3/81, SH., University of York

Social Trends (annually), Central Statistical Office, HMSO. The year quoted in the reference is the year of the survey, rather than the year of publication.

Sonderegger, R. C. (1978), 'Movers and stayers: the residents' contribution to variation across houses in energy consumption in space heating', in R. H. Socolow (ed.), *Saving energy in the home*, Ballinger, Cambridge, Mass.

South London Consortium (1989), *Insulation and associated works 1989-1991*, London Boroughs Bulk Quotation, SLC, London

SPRU (1988), *ESRC designated research centre for science, technology and energy policy: report for mid-term review*, Part II, Science Policy Research Unit, University of Sussex, Brighton

SSS (annually), *Social Security Statistics*, Department of Health and Social Security, HMSO. The year quoted in the reference is the year of the survey, rather than of publication.

Stitt, D. (1990), *Something has to go to the wall*, Right to Fuel Northern Ireland, Belfast

Strauss, L. (1954), Remarks prepared by Lewis L. Strauss, Chairman, United States Atomic Energy Commission, for delivery at the Founders' Day Dinner, National Association of Science Writers, on Thursday, 16 September, 1954, New York

Surrey, J. and Chesshire, J. (1979), 'Energy policy and energy strategy', paper for the Consumer Representatives on the Energy Commission, SPRU, University of Sussex, Brighton

Sutherland, C. M. J. (1990), *Comparative domestic heating cost tables*, Sutherland Associates, Banstead, Surrey

Townsend, P. (1979), *Poverty in the United Kingdom: a survey of household resources and standards of living*, Penguin, Harmondsworth, Middx.

Townsend, P. (1990), 'And the walls come tumbling down), *Poverty*, CPAG Magazine, no. 75, Spring, pp.8-11

Uglow, C. (1981), 'The calculation of energy use in dwellings', *Building Services Engineering and Technology*, vol. 2, no. 1, pp.1-14

UN (1980), *Demographic yearbook*, United Nations

UN (1985), *Demographic yearbook*, United Nations

Watt Committee on Energy (1979), *A warmer house at lower cost*, London

Wheeler, R. (1985), *Staying put — a research project into the growing problem of elderly owner-occupiers*, Building Societies Association, London

WHO (1987), *Health impact of low indoor temperatures*, World Health Organization, Copenhagen

Wicks, M. (1978), *Old and cold: hypothermia and social policy*, Heinemann

Williams, D. (1986), 'Heating homes — the human factor', paper presented to *Unhealthy Housing Conference*, University of Warwick, Institution of Environmental Health Officers, London, 14-16 December

Williams, F. (1977), 'Introduction' in F. Williams (ed.), *Why the poor pay more*, Macmillan, London

Winfield, M. (1982), *The human cost of fuel disconnections*, discussion paper, Family Service Units, London

Woollcombe, J. (1938), *Houses: advice to those about to buy or build*, Clay Products Technical Bureau of Great Britain, London

Wright, L. (1964), *Home fires burning: the history of domestic heating and cooking*, Routledge & Kegan Paul, London

Wynn, M. and Wynn, A. (1979), *Prevention of handicap and the health of women*, Routledge & Kegan Paul, London

Index